RESEARCH REPORT

Conflict Projections in U.S. Central Command

INCORPORATING CLIMATE CHANGE

Mark Toukan, Stephen Watts, Emily Allendorf, Jeffrey Martini,
Karen M. Sudkamp, Nathan Chandler, Maggie Habib

For more information on this publication, visit **www.rand.org/t/RRA2338-3**.

About RAND

The RAND Corporation is a research organization that develops solutions to public policy challenges to help make communities throughout the world safer and more secure, healthier and more prosperous. RAND is nonprofit, nonpartisan, and committed to the public interest. To learn more about RAND, visit www.rand.org.

Research Integrity

Our mission to help improve policy and decisionmaking through research and analysis is enabled through our core values of quality and objectivity and our unwavering commitment to the highest level of integrity and ethical behavior. To help ensure our research and analysis are rigorous, objective, and nonpartisan, we subject our research publications to a robust and exacting quality-assurance process; avoid both the appearance and reality of financial and other conflicts of interest through staff training, project screening, and a policy of mandatory disclosure; and pursue transparency in our research engagements through our commitment to the open publication of our research findings and recommendations, disclosure of the source of funding of published research, and policies to ensure intellectual independence. For more information, visit www.rand.org/about/principles.

RAND's publications do not necessarily reflect the opinions of its research clients and sponsors.

Published by the RAND Corporation, Santa Monica, Calif.

Library of Congress Cataloging-in-Publication Data is available for this publication.
ISBN: 978-1-9774-1246-1

ABOUT THIS REPORT

THIS REPORT PRESENTS an analysis of the impact of climate change on the frequency of conflict in the U.S. Central Command (CENTCOM) area of responsibility (AOR). In our analysis, we characterize the current state of scholarship on the topic before applying a machine learning framework to generate new conflict projections. We then identify reasons why existing work—including our own—might underestimate the relationship between climate hazards and conflict and conclude with an excursion on our modeling that illustrates the potential risk of an unexpected increase in conflict in the theater.

The primary audience for this research is CENTCOM leadership and planners and personnel from CENTCOM's interagency partners (e.g., the U.S. Agency for International Development and U.S. State Department). This report is the third in a series stemming from a larger project to consider the impacts of climate change on the security environment in the region. The first report, *A Hotter and Drier Future Ahead: An Assessment of Climate Change in U.S. Central Command*, presents an analysis of projected climate impacts in the CENTCOM AOR in 2035, 2050, and 2070. The second report, *Pathways from Climate Change to Conflict in U.S. Central Command*, details how climate hazards can manifest in conflict but does not address the frequency with which that conflict could occur. This report addresses that gap.

The fourth report, *Mischief, Malevolence, or Indifference? How Competitors and Adversaries Could Exploit Climate-Related Conflict in the U.S. Central Command Area of Responsibility*, presents an analysis of how competitors—China, Russia, and Iran—might attempt to exploit climate-induced conflict in the CENTCOM AOR. And the final report, *Defense Planning Implications of Climate Change for U.S. Central Command*, examines off-ramps to climate-influenced conflict and the operations, activities, and investments CENTCOM needs to be prepared to execute, given climate impacts on the security environment.

RAND National Security Research Division

This research was sponsored by CENTCOM and conducted within the International Security and Defense Policy Program of the RAND National Security Research Division (NSRD), which operates the National Defense Research Institute (NDRI), a federally funded research and development center sponsored by the Office of the Secretary of Defense, the Joint Staff, the Unified Combatant Commands, the Navy, the Marine Corps, the defense agencies, and the defense intelligence enterprise.

For more information on the RAND International Security and Defense Policy Program, see www.rand.org/nsrd/isdp or contact the director (contact information is provided on the webpage).

Acknowledgments

The authors thank the RAND Corporation management team who provided feedback on drafts of this report. We also thank internal reviewer Caleb Lucas and external reviewer Jannis Hoch, who both provided insightful comments on the report as part of RAND's quality assurance process. The authors also acknowledge the contributions of Rosa Maria Torres, who assisted with the formatting of this document.

CONTENTS

FIGURES AND TABLES

Figures

Tables

KEY FINDINGS

THIS REPORT STEMS from a larger project that considers the impacts of climate change on the security environment in the U.S. Central Command (CENTCOM) area of responsibility. The project explores how the relationship between climate and conflict should inform CENTCOM's long-term planning and decisionmaking. This report builds on the larger project's regional climate assessment, characterizes the state of research on the relationship between climate and conflict, and presents new analysis that is based on machine learning conflict projections.

- Under all plausible socioeconomic and climate conditions, the CENTCOM area of responsibility will experience substantial conflict in the next half century.
- In areas where climate hazards increase conflict risk, the hazards do so by interacting with other variables that are stronger predictors of conflict.
- Although there is suggestive evidence that worse climate outcomes will correlate with a greater incidence of conflict between 2040 and 2060, temperature increases and declines in precipitation are not the major drivers of the security environment, according to our machine learning model.
- There are good reasons to believe that the existing research and our own conflict forecasts could be underestimating the impact of climate variables on conflict.
- The main limitations of the existing research are inadequate attention paid to
  - the link between climate and conflict, which is neither unidirectional nor direct; rather, the presence of conflict limits a state's ability to adapt to climate change and further increases the risk of conflict; additionally, climate hazards could suppress economic development and contribute to conflict via socioeconomic conditions
  - how climate change could contribute to conditions that shape conflict risk in a manner that is fundamentally different from the conditions that characterized the recent past
  - how climate hazards could—via migration and food price shocks—generate conflict that is far from localized climate impacts or result in conflict in future periods.
- After making assumptions that are grounded in existing research about the impact of drought on the economies of agriculture-dependent areas, we project an increased risk of conflict in those areas.

CHAPTER 1

THE IMPACT OF CLIMATE CHANGE ON CONFLICT

THIS REPORT ADDRESSES how climate change could affect the frequency of conflict in the U.S. Central Command (CENTCOM) area of responsibility (AOR).[1] In the second report in this series, *Pathways from Climate Change to Conflict in U.S. Central Command*, RAND researchers found that climate hazards could contribute to conflict through numerous pathways.[2] In these pathways, climate change affects different types of insecurity (e.g., food, livelihood), which then combine with impacts on state capacity, population flows, and other factors to influence individual and armed group incentives to mobilize and ultimately engage in armed conflict. That research also identified past conflicts in the AOR for which climate hazards were a contributing factor. Building off this prior report, this research addresses the *frequency* of future climate-induced conflict. See Figure 1.1 for the progression of the reports in this series.

We begin this report by examining how the current literature characterizes the relationship between climate change and the incidence of conflict. We then build on that literature to generate conflict projections for the AOR at the provincial level out to 2070. The projections are made on the basis of a machine learning framework that uses historical data to train and validate a forecasting model. The projections incorporate anticipated changes in temperature and levels of precipitation, although these climate factors are used to complement other known drivers of conflict, such as an area's political and economic development.

After we present these conflict projections, we analyze why the consensus in the field and our own modeling could be underestimating the strength of the relationship between climate change and future conflict. To explore what the implications of these potential undercounts are, we present an excursion on our conflict forecasts that accounts for declines in gross domestic product (GDP) growth during droughts in agriculture-dependent areas. The excursion supports the view that the climate-conflict nexus might be underestimated in existing modeling.

The purpose of this research is to support CENTCOM leadership and planners and their interagency partners to prepare for a future security environment that is affected by climate change. Understanding the frequency of future conflict in the AOR, as well as the marginal increase that is owed to climate change as a threat multiplier, will enable the U.S. government to better prepare for this future. A fifth and final report addresses the implications for CENTCOM in terms of the operations, activities, and investments (OAIs) the command could take to mitigate the risk of climate-induced conflict.[3]

The Impact of Climate Change on the Incidence of Conflict

To the extent that there is any consensus among experts about the link between climate change and armed conflict, it is largely captured in the work of the Intergovernmental Panel on Climate Change (IPCC) and the U.S. intelligence community's (IC's) National Intelligence Estimate. These publications are derived from a distillation of existing academic literature on the topic, which generally finds with moderate confidence that climate change will have impacts on some forms of violence but that those impacts will be (1) highly conditional on

Figure 1.1. Progression of Reports in This Series

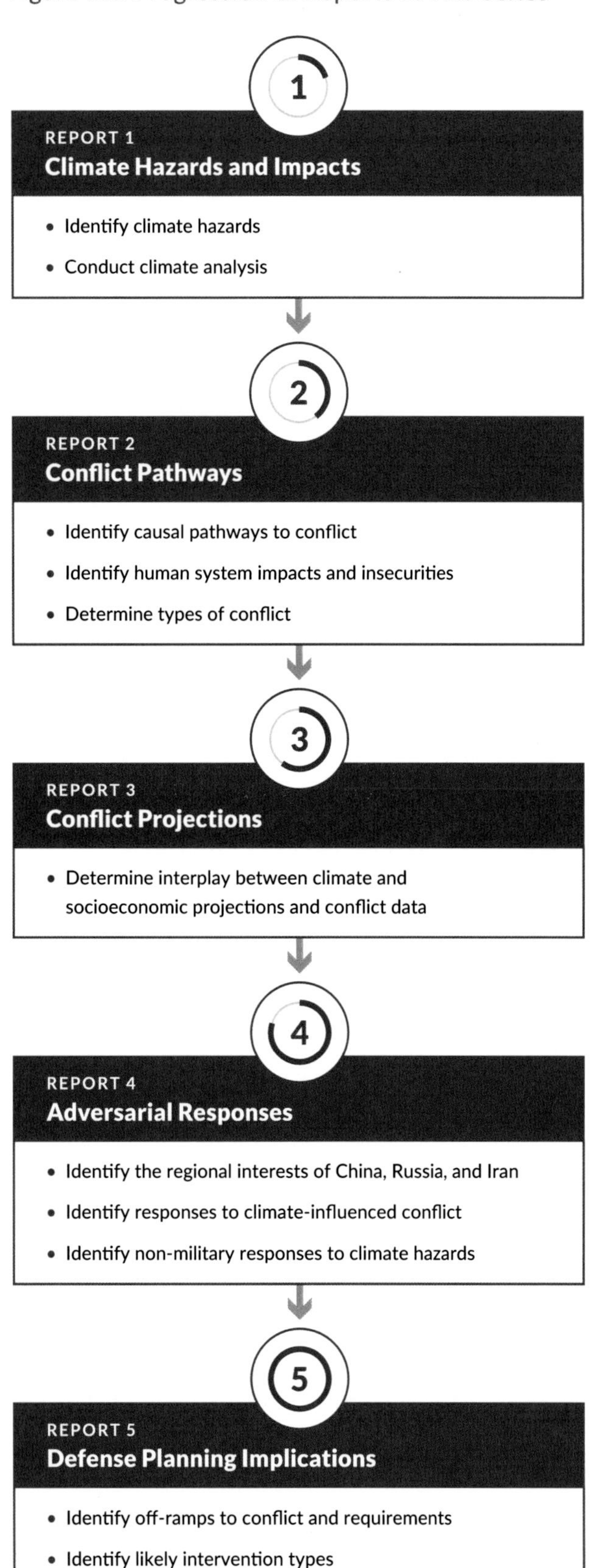

contextual factors and (2) more likely to manifest in instability vice war. After an extensive review of the field in 2022, the IPCC determined that

> [c]limate variability and extremes are associated with increased prevalence of conflict, with more consistent evidence for low-intensity organised violence than for major armed conflict (medium confidence). Compared with other socioeconomic drivers, the link is relatively weak (medium confidence) and conditional on high population size, low socioeconomic development, high political marginalization and high agricultural dependence (medium confidence). Literature also suggests a larger climate-related influence on the dynamics of conflict [such as its duration or intensity] than on the likelihood of initial conflict outbreak (low confidence). There is insufficient evidence at present to attribute armed conflict to climate change.[4]

This view is largely consistent with what was expressed in the 2021 National Intelligence Estimate, which stated that

> intensifying physical effects of climate change out to 2040 and beyond will be most acutely felt in developing countries, which we assess . . . will increase the potential for instability and possibly internal conflict in these countries, in some cases creating additional demands on U.S. diplomatic, economic, humanitarian, and military resources . . . [but the] IC has low to moderate confidence in assessing how climate change effects could cascade in ways that affect U.S. national security interests.[5]

Similarly, the IC's 2023 U.S. annual threat assessment indicated that climate change might affect U.S. interests, although the assessment generally highlighted the potential for elevated tensions rather than predicting a substantial increase in violent conflict:

> Climate change will increasingly exacerbate risks to U.S. national security interests as the physical impacts increase and geopolitical tensions mount about the global response to the challenge. . . . As temperatures rise and more extreme climate effects manifest, there is a growing risk of conflict over resources associated with water, arable land, and the Arctic. . . . Climate-related disasters in low-income countries will deepen economic challenges, raise the risk of inter-communal conflict over scarce resources, and increase the need for humanitarian and financial assistance. The growing gap between the provision of basic needs and what governments and the interna-

tional community can provide raises the likelihood of domestic protests, broader instability, extremist recruitment, and migration.[6]

The authors of these reports chose their words carefully, so it is worth parsing these passages closely. Two areas in particular stand out: (1) conflict, instability, and violence and (2) context and causation.

Conflict, Instability, and Violence

All three reports from the IPCC and the U.S. IC indicate the potential for higher levels of conflict, instability (geopolitical tensions or domestic political unrest), and violence. However, the reports are more cautious about suggesting that these dynamics would rise to the sort of armed conflict that affects U.S. security interests. Academic studies of political violence often use a cutoff of 1,000 deaths in battle in a year to distinguish between low-intensity violence and high-intensity armed conflict.[7] Although this cutoff is artificial, in practice it distinguishes between such events as low-level guerrilla warfare or border skirmishes on the one hand and organized, enduring, high-intensity violence, such as insurgencies or interstate wars, on the other. In general, these official assessments express much more confidence that climate change will be linked to the former type of violence than the latter. The IPCC also appears more skeptical of the potential for a link between climate change and interstate armed conflict.[8]

Context and Causation

The U.S. IC reports suggest that climate change is likely to cause some increase in violence. The IPCC report is more circumspect. It indicates that developing countries are particularly exposed to the effects of climate change and notes that such environments are at an elevated risk of violence. But the IPCC report stops short of claiming that climate change will drive increases in the incidence of armed conflict. Thus, many developing country contexts are vulnerable to both the effects of climate change and violence without a causal connection between the two.

The Interaction of Climate Change with Other Variables

An important takeaway from the judgments of the international community and U.S. IC—as well as from our prior report on causal pathways from climate hazards to conflict—is that policymakers should view climate change as interacting with other factors that predispose an area to conflict.[9] Some of the most prominently cited exacerbating factors and mitigating circumstances are presented in Table 1.1. Simply put, certain socioeconomic, political, security, and demographic conditions interact with climate hazards to increase or decrease (mitigate) the climate-conflict relationship.[10] This means that finding a tangible climate-conflict link is highly conditional on a variety of intervening factors. While a complete accounting of the factors addressed in the literature is beyond the scope of this analysis, in Chapter 2 of this report, we examine the interaction between climate hazards and these scope conditions to model future conflict in the CENTCOM AOR.

Table 1.1. Scope Conditions That Make Climate Hazards More or Less Likely to Manifest in Conflict

Factor	More Likely to Generate Conflict	Less Likely to Generate Conflict
Regime type	Hybrid regimes	Fully consolidated democratic or authoritarian regimes
Economic growth	Low and volatile economic growth	Consistently strong economic growth
Conflict history	Preexisting conflict or neighboring an area in conflict	Distant and infrequent experience with prior conflict; removed from regional instability
Population size	Large populations that lead to more opportunity for intergroup conflict; could strain government capacity	Smaller populations that are associated with less intergroup conflict

SOURCES: Nils Petter Gleditsch, "This Time Is Different! Or Is It? NeoMalthusians and Environmental Optimists in the Age of Climate Change," *Journal of Peace Research*, Vol. 58, No. 1, January 2021; Håvard Hegre, Halvard Buhaug, Katherine V. Calvin, Jonas Nordkvelle, Stephanie T. Waldhoff, and Elisabeth Gilmore, "Forecasting Civil Conflict Along the Shared Socioeconomic Pathways," *Environmental Research Letters*, Vol. 11, No. 5, April 2016; Frank D. W. Witmer, Andrew M. Linke, John O'Loughlin, Andrew Gettelman, and Arlene Laing, "Subnational Violent Conflict Forecasts for Sub-Saharan Africa, 2015–65, Using Climate-Sensitive Models," *Journal of Peace Research*, Vol. 54, No. 2, March 2017; Benjamin T. Jones, Eleonora Mattiacci, and Bear F. Braumoeller, "Food Scarcity and State Vulnerability: Unpacking the Link Between Climate Variability and Violent Unrest," *Journal of Peace Research*, Vol. 54, No. 3, May 2017.

Endnotes

[1] The CENTCOM AOR includes 21 states and encompasses Egypt, the Levant, the Arabian Peninsula, Iran, and South and Central Asia. Israel was added to the combatant command in 2021 per changes in the unified command plan.

[2] Nathan Chandler, Jeffrey Martini, Karen M. Sudkamp, Maggie Habib, Benjamin J. Sacks, and Zohan Tariq, *Pathways from Climate Change to Conflict in U.S. Central Command*, RAND Corporation, RR-A2338-2, 2023.

[3] Karen M. Sudkamp, Elisa Yoshiara, Jeffrey Martini, Mohammad Ahmadi, Matthew Kubasak, Alexander Noyes, Alexandra Stark, Zohan Hasan Tariq, Ryan Haberman, and Erik E. Mueller, *Defense Planning Implications of Climate Change for U.S. Central Command*, RAND Corporation, RR-A2338-5, 2023.

[4] Hans-Otto Pörtner, Debra C. Roberts, Melinda M. B. Tignor, Elvira Poloczanska, Katja Mintenbeck, Andrés Alegría, Marlies Craig, Stefanie Langsdorf, Sina Löschke, Vincent Möller, Andrew Okem, and Bardhyl Rama, eds., *Climate Change 2022: Impacts, Adaptation and Vulnerability, Contribution of Working Group II to the Sixth Assessment Report of the Intergovernmental Panel on Climate Change*, Intergovernmental Panel on Climate Change, Cambridge University Press, 2022, p. 2414. For a comparative textual analysis of the IPCC's uncertain and evolving views on the climate-conflict link across its Third (2001), Fourth (2007), and Fifth (2014) Assessment Reports, see Nils Petter Gleditsch and Ragnhild Nordås, "Conflicting Messages? The IPCC on Conflict and Human Security," *Political Geography*, Vol. 43, November 2014.

[5] National Intelligence Council, *National Intelligence Estimate: Climate Change and International Responses Increasing Challenges to US National Security Through 2040*, NIC-NIE-2021-10030-A, 2021, pp. i–ii.

[6] Office of the Director of National Intelligence, *Annual Threat Assessment of the U.S. Intelligence Community*, February 6, 2023, p. 22.

[7] For example, a high-intensity conflict would include the ongoing civil war in Syria or in the Democratic Republic of the Congo from 1998 to 2003.

[8] IPCC, 2022, pp. 1087–1088.

[9] Chandler et al., 2023.

[10] As Koubi concludes, "Substantial agreement exists that climatic changes contribute to conflict under some conditions and through some pathways. . . . Future research should continue to investigate how climatic changes interact with and/or are conditioned by socioeconomic, political, and demographic settings to cause conflict and uncover the causal mechanisms that link these two phenomena" (Vally Koubi, "Climate Change and Conflict," *Annual Review of Political Science*, Vol. 22, May 2019, p. 343).

CHAPTER 2

FORECASTS OF FUTURE INTERNAL CONFLICT

THIS CHAPTER PRESENTS forecasts of intrastate conflict in the CENTCOM AOR from 2030 to 2070 using a machine learning forecasting framework. The model uses scenario-based projections of climate indicators and socioeconomic factors to identify plausible future trends in civil conflict across the region. As discussed in Chapter 1, there is only low to moderate confidence on how such climate factors as extreme heat and drought affect conflict. However, climate change will affect a variety of factors for which we have greater certainty of the additional conflict risk they will introduce. Those factors include population size and distribution, economic development, governance, and prior exposure to conflict. Our forecasts integrate both sets of variables: the climate variables with highly uncertain impacts on conflict and the underlying socioeconomic, political, and prior conflict variables with more predictable impacts on conflict.

Machine learning approaches are appropriate for making forecasts under conditions in which it is difficult to know beforehand the shape of the underlying relationships. We begin with a brief literature review of the existing climate-conflict forecasting work. We then present the machine learning framework and data sources that we used to produce conflict forecasts. Finally, we describe how projected conflict risks change for the CENTCOM AOR across plausible climate change and adaptation scenarios.

Existing Work

The existing research that investigates the connection between climate and conflict covers a variety of socioeconomic and climate predictors. Geographically, Africa is a common region of study. In their forecasts of conflict in Africa, Hoch and colleagues find that the changes in hydroclimatic variables are predictive of greater or lesser conflict depending on the location. However, they find climate variables to be less predictive than variables related to prior conflict, quality of governance, education, and population.[1] Likewise, Witmer and colleagues find no significant relationship for precipitation anomalies and a weak but significant effect of temperature anomalies in forecasting conflict in Sub-Saharan Africa out to 2065.[2] Poor governance and high fertility rates are the most important predictors in their forecasts. Hegre and colleagues forecast global conflict using the Shared Socioeconomic Pathways (SSP) framework (discussed in the following paragraphs). They do not find temperature to be a significant factor in their historical model, and temperature does not improve the predictive performance of their forecasts.[3]

Linke, Witmer, and O'Loughlin take a disaggregated approach to conflict prediction in Kenya by using survey data from reported instances of conflict alongside such climate variables as drought frequency, weather variability, and vegetation health.[4] Models that included the climate variables performed worse than models that used only demographic and contextual predictors. Harari and LaFerrara forecast civil conflict in Africa using the Standardized Precipitation Evapotranspiration Index.[5] They find that drought shocks during the growing season are associated with increased conflict risk in the future and that conflict risk tends to spread to neighboring areas. However, drought shocks and conflict are only related during the growing season, and drought shocks were not significantly related to conflict in neighboring cells, suggesting that the effects of drought shocks stay local.

As these studies suggest, the nature and strength of the relationship between climatic variables and conflict is likely to depend heavily on the socioeconomic context in which climate change plays out. While there are inherent limitations to conflict forecasts, such as the inability to account for relationships between risk factors and conflict that are not reflected in historical data, the use of these methods alongside scenario-based projections can still prove valuable for understanding plausible futures and how different social and policy pathways might influence the future security landscape.[6] The rest of this chapter presents the application of a machine learning framework to make the first scenario-based forecast of climate and conflict in the CENTCOM AOR.

Data and Methods

CoPro Forecasting Framework and Conflict Data

We use an open-source conflict forecasting framework, CoPro, which has been used previously to forecast how climate change could affect the risk of internal conflict in Africa.[7] The framework is built to allow users to specify climate and other indicators for forecasting civil conflict in specified geographic areas using the Uppsala Conflict Data Program's Georeferenced Event Dataset (GED).[8] These data record the location of internal violent conflicts with latitude and longitude. Along with location data, GED records the time, intensity in terms of fatalities, and type of conflict.[9]

There are three types of conflict covered by GED: (1) state-based violent conflict, which is conflict between the government of a state and one or more organized non-state groups over the control of government or territory; (2) one-sided violence, which is violence against civilians by the government or another organized group; and (3) non-state conflict, which is violent conflict between organized groups that does not involve the government as a belligerent.[10] The first type of conflict includes intrastate war: organized conflict over the state. GED collects battle death data at the individual level, so this category contains violence above and below the threshold of war, which is typically 25 battle-related deaths. The second and third categories reflect violence that is not typically considered intrastate war, such as violent state repression or communal violence.

CENTCOM planners and other readers with practical applications in mind should focus on two implications from the data inputs. First, GED does not cover interstate war, which remains a much rarer event than civil conflict; therefore, interstate war is not modeled in our analysis. Second, when we report the findings from our modeling as the overall frequency of conflict, the reader should keep in mind that these conflicts also include lower-level violence—between 25 and 999 annual battle deaths—that is generally not regarded as full-fledged war.

Modeling Process

CoPro trains and tests a user-specified machine learning model on historical data.[11] We average across ten model runs to increase the robustness of the results. We train the model on data from 1995 through 2015, the most recent year for which we have historical data for all indicators included in the model. Each model is trained on 70 percent of the historical data and tested on the remaining 30 percent, where each set of training and test data is randomly selected across separate model runs. (See Figure 2.1 for a visual depiction of the process.)

We use random forest models to produce conflict projections. The random forest algorithm creates decision trees by randomly selecting an explanatory factor, such as temperature or population, to create a branch. The algorithm groups similar observations together when selecting a value to determine which observations go on which side of a branch. For example, provinces with higher wealth that are less prone to civil conflict will tend to be grouped down the same branch of a tree. As the algorithm proceeds, it makes additional branches with randomly selected factors. For example, a subsequent split might sort provinces by governance score. This process yields similar groupings of provinces at the end of each tree. By iterating this process across many trees and randomly selecting factors to sort cases, the algorithm can classify provinces in terms of similar conflict risk.[12] But like other techniques for analyzing the relationship between certain factors and an outcome, this machine learning approach determines the nature of the relationship by extrapolating from the past. Thus, the model would not predict a future shift in the relationship between climate factors and conflict outcomes that has yet to appear.

Geographic Extent and Data Aggregation

The modeling framework aggregates all data and predictions by province. Because our application of the model focuses on civil conflict, which often plays out differently at the local level, we report out our results by province.[13] For climate indicators, the model takes the average value across all grid cells contained in the province. We take the natural logarithm of

Figure 2.1. Machine Learning Approach

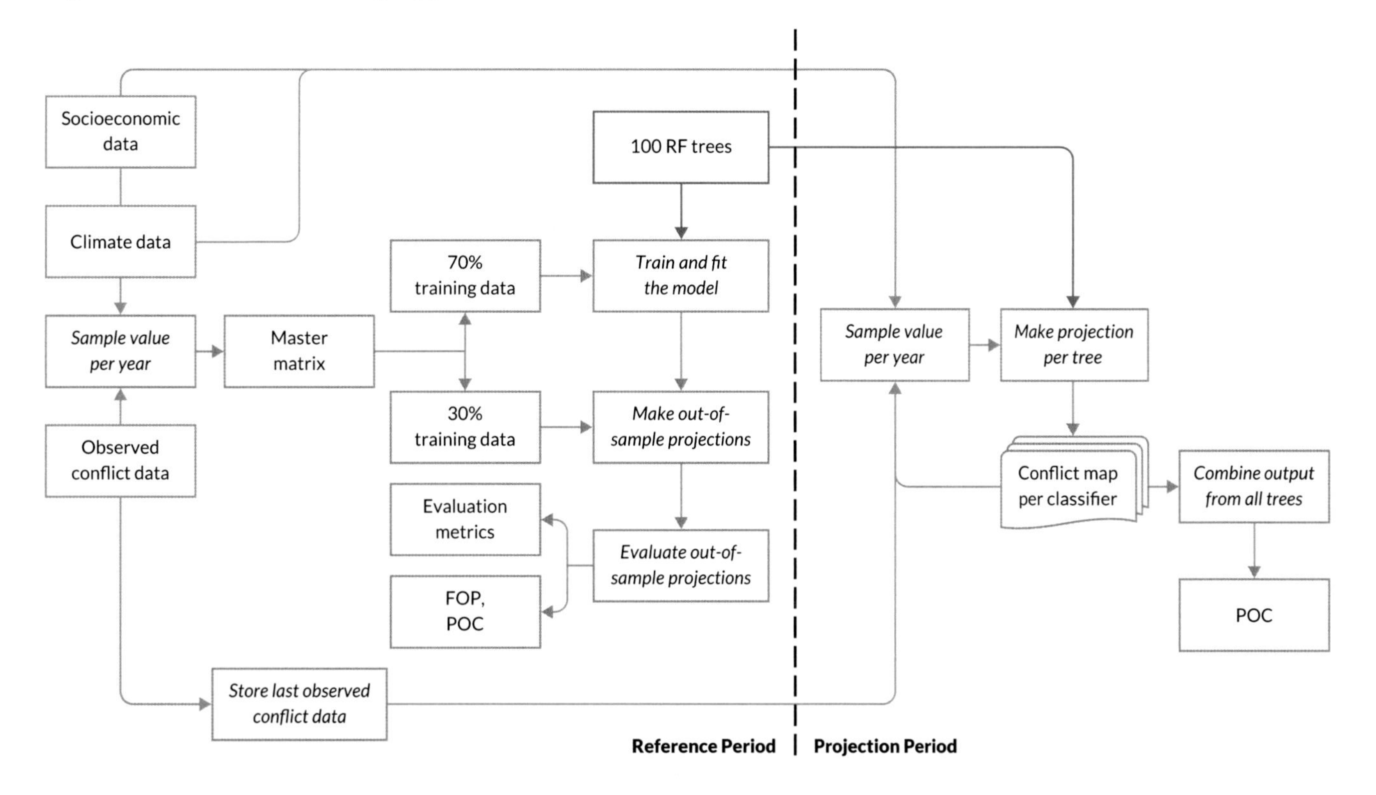

SOURCE: Adapted from Hoch, de Bruin, and Wanders, 2021.

NOTE: FOP = fraction of correct predications; POC = probability of conflict; RF = random forest.

the sum of population and GDP per capita by province and the average governance indicator by province.

Climate Data

We draw forecasts of climate indicators from climate models from the sixth iteration of the Coupled Model Intercomparison Project (CMIP6).[14] We include indicators of the average precipitation and average temperature.[15] These indicators are constructed from datasets from the Inter-Sectoral Impact Model Intercomparison Project (ISIMIP) and are available in a 0.5 × 0.5-degree grid of roughly 55 kilometers.[16]

We use historical climate indicators from 1995 through 2015 and forecast indicators for 2016–2050 for two climate scenarios, Representative Concentration Pathways (RCPs) 6.0 and 8.5, which are indicative of climate changes resulting from moderate- and high-emission scenarios and greenhouse gas concentration scenarios, respectively.[17] Relative to a preindustrial baseline, RCP 6.0 projects a temperature increase of 4 to 6.5°F over the last two decades of the 21st century (2081–2100), whereas RCP 8.5 projects a global average temperature increase of 5.4 to 11.2°F at the same points in time.[18]

For an accessible presentation of how these climate changes will affect the physical environment in CENTCOM, please see the first report in this series, *A Hotter and Drier Future Ahead: An Assessment of Climate Change in U.S. Central Command*.[19] The first report finds that the AOR will grapple with escalating issues of high temperature, drought, and long-term dryness. The increasing aridity will deplete existing water resources, intensifying regional tensions because of preexisting water scarcity. Despite longer growing seasons in some areas, such as Central Asia, because of warming, these harsher conditions will challenge agricultural activity across the AOR. Furthermore, countries such as Yemen, Oman, and Pakistan could simultaneously experience aridification and heightened extreme precipitation, elevating risks of flash flooding.

Non-Climate Data

We account for conflict in a province in the previous year and for conflict that occurs in neighboring grid cells.[20] Civil conflict tends to reoccur or spread to neighboring states, so it is important to account for the tendency of geographic areas to enter into conflict traps that persist over time.

We present forecasts that track different climate scenarios and plausible societal responses to those scenarios. Climate researchers have developed a set of SSPs that present narratives of different challenges that societies face in mitigating and adapting to climate change.[21] These narratives have served as a basis for the research community to produce forecasts of important socioeconomic indicators that are consistent with each SSP. We include forecasts of population and GDP, both collected at the level of the 0.5 × 0.5-degree grid cell, and countrywide governance indicators for three SSPs depicted in Figure 2.2 by their defining characteristics.[22]

We selected these SSPs to be consistent with the RCP scenarios used for the climate forecasts—scenarios that reflect at least moderate challenges in mitigating climate effects. Note that SSP 5 is the only socioeconomic scenario with assumptions that are consistent with both RCPs 6.0 and 8.5.[23] This is important because one way to see the effect of climate in conflict forecasts is to hold the socioeconomic conditions constant (i.e., use SSP 5) while varying the climate scenarios (i.e., comparing the results of RCP 6.0 and 8.5) This is the only combination (SSP 5/RCP 6.0 vs. SSP 5/RCP 8.5) for which we can isolate the impact of climate change on conflict forecasts with all other conditions held constant.

The conditions contained in the SSPs, listed in Table 2.1, are consistently associated with conflict risk. Provinces with a high population are more likely to experience civil conflict.[24] All else equal, more-populated areas are likely to contain a greater mix of sociopolitical groups that might come into conflict. Similarly, larger populations provide a larger pool of individuals who might participate in conflict. Individuals in more-developed provinces face higher opportunity costs of conflict, face fewer grievances on average than in less-developed provinces, and are more likely to be deterred by high-capacity states. Finally, states with high-quality governance are less likely to experience conflict. Better-governed states are less likely to provide the impetus for rebellion by delivering public goods and are more likely to have institutions that do not discriminate on the basis of group identity. Similarly, states with stronger governance can prevent conflict between groups by providing venues for nonviolent conflict resolution and by providing public goods that reduce intergroup tensions.

Results

We evaluate the forecasting model using data from 1995 through 2015. Figure 2.3 and similar figures are based on our analysis of data from the CoPro forecasting framework. Figure 2.3 displays a map that shows the accuracy of the model in predicting conflict in the historical reference data.

Figure 2.2. Shared Socioeconomic Pathways by Defining Challenges

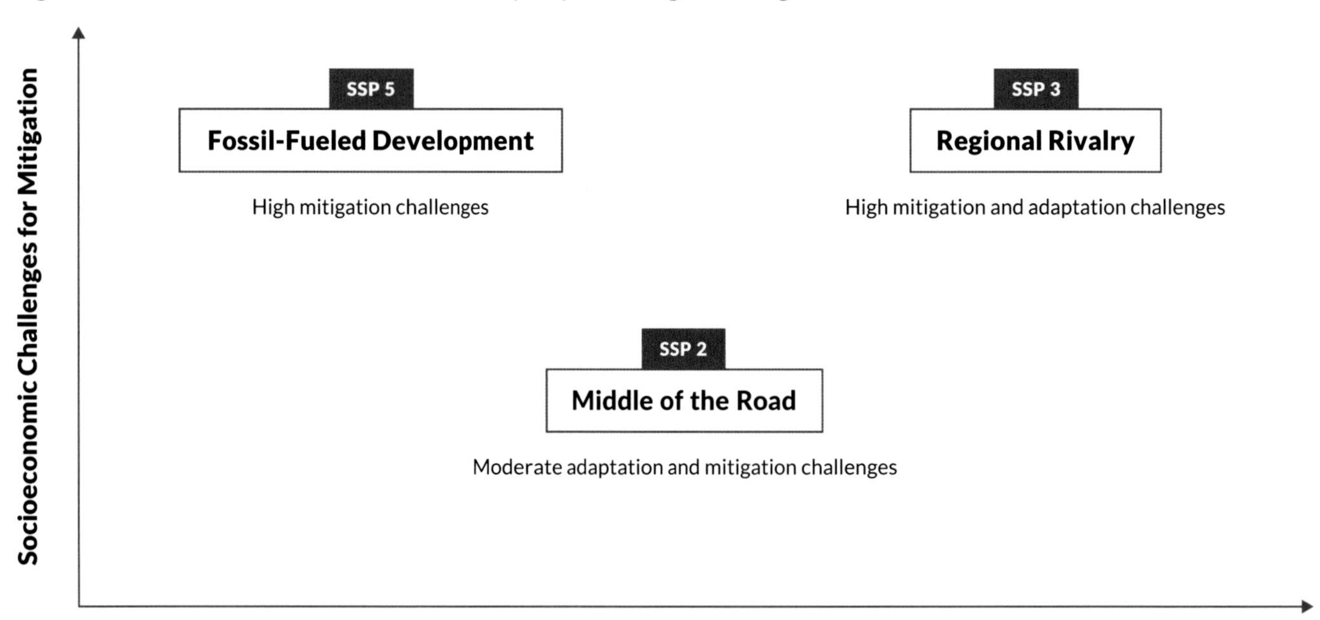

SOURCE: Adapted from Keywan Riahi, Detlef P. van Vuuren, Elmar Kriegler, and Brian O'Neill, "The Shared Socio-Economic Pathways (SSPs): An Overview," poster, International Committee on New Integrated Climate Change Assessment Scenarios, undated.

Table 2.1. Shared Socioeconomic Pathways Characteristics

	Narrative	Economic Development in 2100[a]	Population in 2100	Urbanization in 2100
SSP 2	*Middle of the road:* The world follows a path along which social, economic, and technological trends do not shift markedly from historical patterns.	The global average of GDP per capita eclipses $50,000.	The global population dips below replacement level around 2060, bringing the total population to roughly 9 billion in 2100.	Approximately 80% of the global population is urbanized.
SSP 3	*Regional rivalry—a rocky road:* Resurgent nationalism, concerns about competitiveness and security, and regional conflicts push countries to increasingly focus on domestic or, at most, regional issues.	The global average of GDP per capita grows incrementally to $20,000.	The global population reaches 12 billion in 2100.	The slowest urbanization rate of all the SSPs, with just under 60% of the global population urbanized in 2100.
SSP 5	*Fossil-fueled development—taking the highway:* This world places increasing faith in competitive markets, innovation, and participatory societies to produce rapid technological progress and the development of human capital as the path to sustainable development.	The global average of GDP per capita reaches nearly $120,000.	The global population dips below replacement level around 2050, bringing the total population to roughly 7 billion in 2100.	Approximately 90% of the global population is urbanized.

SOURCE: Features information from Keywan Riahi, Detlef P. van Vuuren, Elmar Kriegler, Jae Edmonds, Brian C. O'Neill, Shinichiro Fujimori, Nico Bauer, Katherine Calvin, Rob Dellink, Oliver Fricko, Wolfgang Lutz, Alexander Popp, Jesus Crespo Cuaresma, Samir K. C., Marian Leimbach, Leiwen Jiang, Tom Kram, Shilpa Rao, Johannes Emmerling, Kristie Ebi, Tomoko Hasegawa, Petr Havlik, Florian Humpenöder, Lara Aleluia Da Silva, Steve Smith, Elke Stehfest, Valentina Bosetti, Jiyong Eom, David Gernaat, Toshihiko Masui, Joeri Rogelj, Jessica Strefler, Laurent Drouet, Volker Krey, Gunnar Luderer, Mathijs Harmsen, Kiyoshi Takahashi, Lavinia Baumstark, Jonathan C. Doelman, Mikiko Kainuma, Zbigniew Klimont, Giacomo Marangoni, Hermann Lotze-Campen, Michael Obersteiner, Andrzej Tabeau, and Massimo Tavoni, "The Shared Socioeconomic Pathways and Their Energy, Land Use, and Greenhouse Gas Emissions Implications: An Overview," *Global Environmental Change*, Vol. 42, January 2017.

[a] In 2005 U.S. dollars.

Each province is colored according to the fraction of the model's predictions—either of conflict or its absence—that are correct. The model tends to satisfactorily predict which provinces experience a high degree of conflict or very little conflict.[25] For example, much of Central Asia or Oman experiences little conflict, while parts of Yemen have experienced a great deal of conflict. The model performs more modestly predicting conflict in eastern Iran, much of Iraq and Afghanistan, Saudi Arabia, and eastern Pakistan. While overall performance of the model is good, with an average receiver operating characteristic (ROC-AUC) score of 0.92, the figure demonstrates that model projections must be interpreted with caution given the uneven accuracy across the AOR.[26]

Relative Importance of Climate, Conflict, and Socioeconomic Factors

The random forest algorithm allows for comparison of the relative importance of the different factors included in the model. Figure 2.4 displays each of the factors that are included in the model, colored according to whether they are related to conflict (conflict in the prior year, conflict in a neighboring province in the prior year), climate (temperature, precipitation), or socioeconomic factors (GDP, population, governance). These factors are sorted by their permutation importance, which is a metric that captures how much less accurate a model is if random noise is added to that factor. Factors that aid in model prediction will tend to result in lower prediction accuracy if they are randomly changed; these are assigned higher permutation scores to reflect their importance.

The figure identifies governance quality and prior year conflict in a neighboring grid cell as the most important factors. Climate factors (temperature, precipitation) are roughly as important in the model as two socioeconomic factors. It is important to note that we lack data on factors that are known to be important drivers of conflict for each of the climate and socioeconomic scenarios. This adds an additional limitation to the model's ability to accurately forecast conflict. The projections should therefore be interpreted as high-level outcomes that are consistent with a model that includes only a subset of relevant indicators and reflects a specific set of relationships between climatic and socioeconomic indicators.

Figure 2.3. Fraction of Correct Conflict Predictions

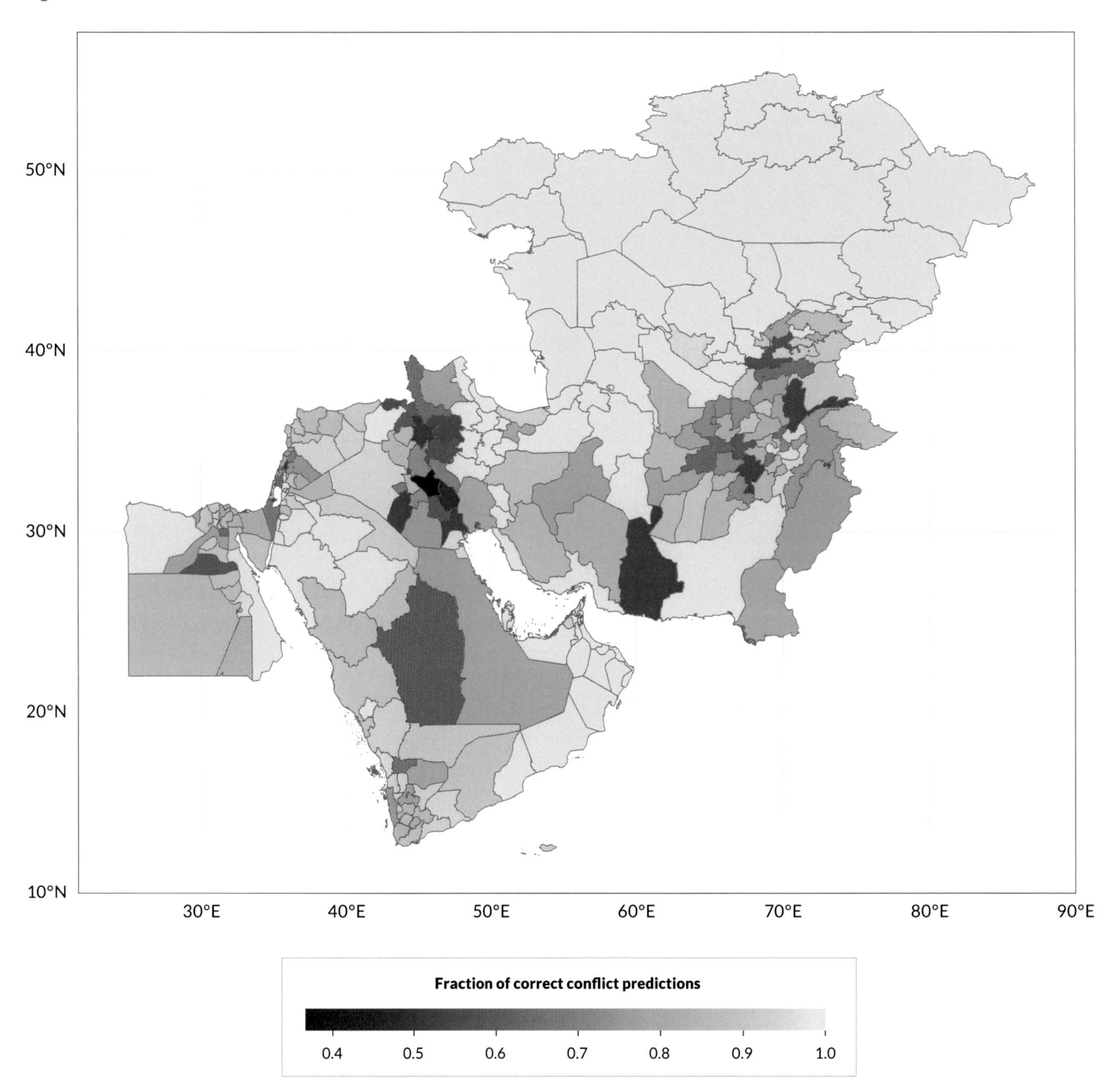

Figure 2.4. Permutation Importance Scores for the Representative Concentration Pathways–Shared Socioeconomic Pathways Model

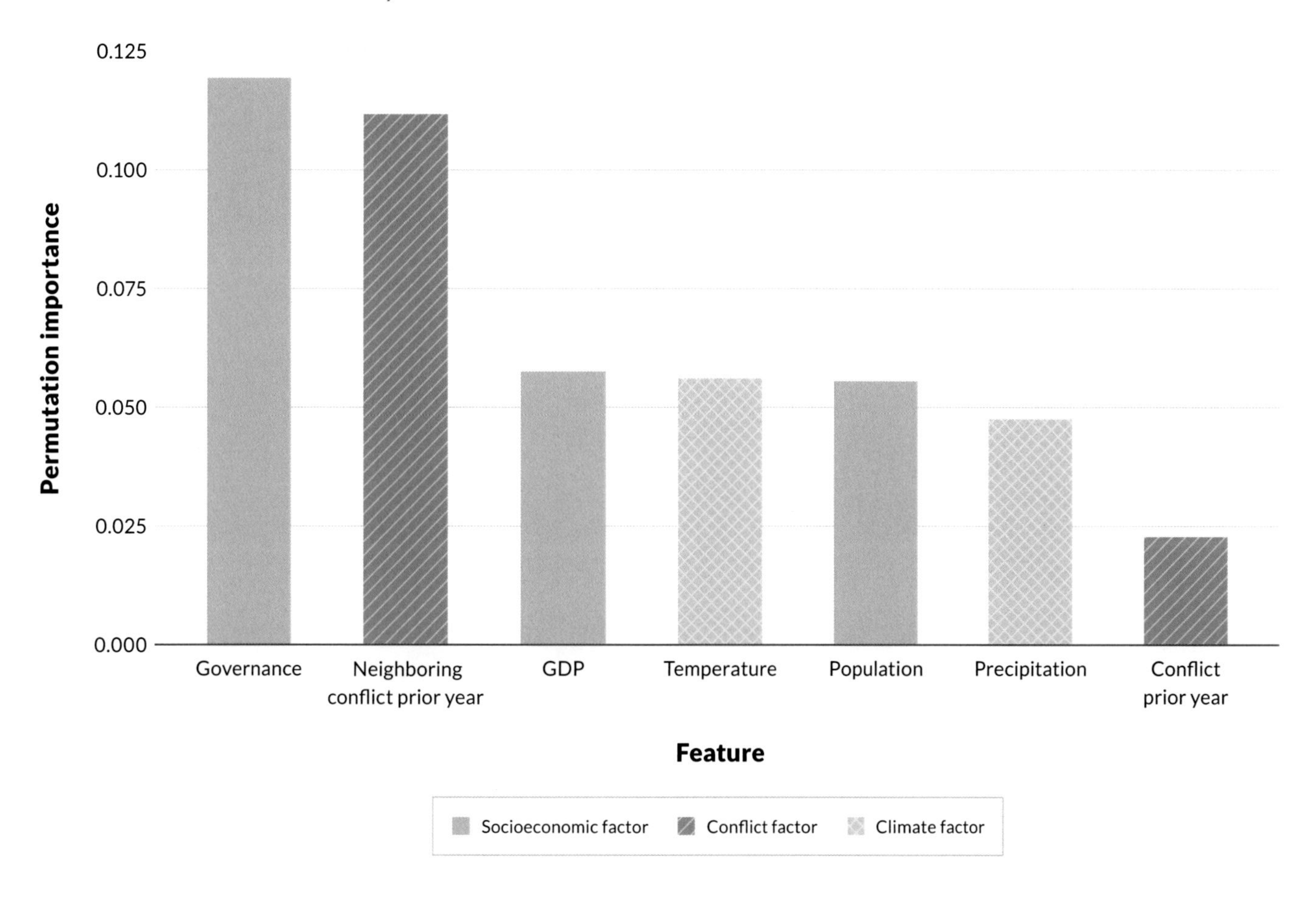

Figure 2.5 displays the conflict probability across the AOR for the portion of the data that was held aside for validation for the reference period 1995–2015. The model predicts a high risk of conflict in Israel, Iraq, Afghanistan, and Pakistan and, to a lesser extent, in Yemen, Syria, and parts of western Iran. These results inform the forecasts across scenarios—the model tends to expect conflict to continue in places where it has historically occurred—and help frame how projections across scenarios differ from the historical baseline.

Scenario Projections

We now turn to projections for the scenarios. To facilitate comparison across scenarios, we present the results first by characterizing average conflict risk across three decades (2040–2070) and the four scenarios (RCP 6.0-SSP 2/3/5 and RCP 8.5-SSP 5) in the forecast. The key in Figure 2.6 denotes the risk of any civil conflict (state-based, one-sided, or non-state) occurring within provincial borders during the indicated decade. A conflict is counted when it generates at least 25 battle deaths. The purple sections of the maps indicate that the risk of civil conflict occurring is 20 percent or higher, whereas the light yellow sections of the map indicate that the risk of such conflict is lower than 5 percent over the decade.

Relative to the baseline probabilities from the reference period depicted in Figure 2.5, a high risk of conflict is expected in many of the same areas across scenarios, especially Iraq, Yemen, and Pakistan. A lower projected risk relative to the baseline is expected in other areas, such as large parts of Afghanistan and Israel, while a higher risk is expected in such areas as Iran across multiple scenarios. It is notable that projections of conflict risk in the northern part of the AOR do not vary greatly across different scenarios. This is likely because of the relative scarcity of civil conflict in Central Asia compared with other parts of the AOR. As shown in Figure 2.5, Tajikistan is the only country with provinces that have a significant probability of conflict in the reference period; Tajikistan also shows greater variability in projected risk across scenarios.

Figure 2.5. Baseline Conflict Probability in Reference Period 1995–2015

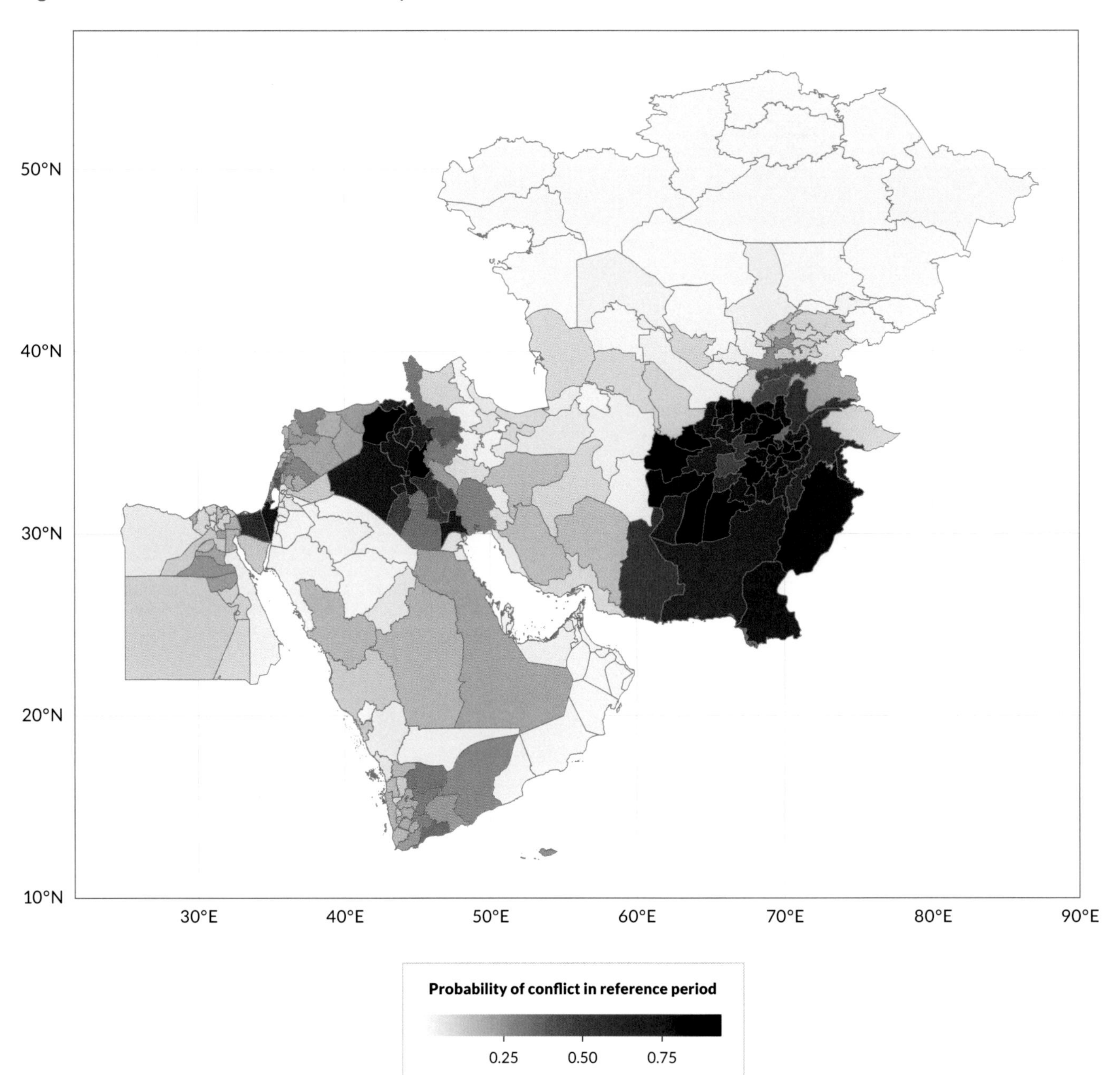

How Projected Conflict Risk Changes Across Scenarios

Comparison of Risk Across Representative Concentration Pathway-Shared Socioeconomic Pathway Scenarios

Figure 2.6 shows that projected conflict risk is highest in the most pessimistic adaptation scenario, RCP 6.0-SSP 3, in which emissions remain high and states face challenges accessing institutional and technological mitigations for adverse climate effects.[27] States simultaneously face large population pressures and increasing inequality in this scenario.[28] This scenario projects the highest conflict risk across all three decades. Projected conflict hot spots through the end of the analysis include large parts of Pakistan, Iraq, Yemen, and Afghanistan. A moderately high risk of conflict is projected for Iran in this scenario. What is notable for CENTCOM is that RCP 6.0 is the current trajectory many climate scientists believe the world is heading toward, and the SSP narrative of resurgent nationalism that coincides with regional conflict will likely strike many military planners as highly plausible.[29]

The greater the ability of states to mitigate and adapt to climate effects internally and to find means of cooperating externally lead to lower projected conflict risk in the RCP 6.0-SSP 2 scenario. This scenario is moderate both in terms of climate trends and in terms of the adequacy of socioeconomic reactions to those trends. Economic development in this scenario is higher than in SSP 3 and population growth lower, although states still face moderate challenges with inequality and climate change mitigation. Projected conflict in this scenario is lower than in the RCP 6.0-SSP 3 scenario for each decade of the projection. Much of Pakistan, however, is still forecast to have a high risk of conflict.

SSP 5 assumes high economic growth driven by fossil fuel consumption. Under this scenario, population growth is low, in accordance with rapid economic development, and inequalities less pronounced given investments in human capital.[30] The two RCP 5 scenarios project overall conflict risk lower than in RCP 6.0-SSP 3. As Figure 2.6 suggests, SSP 5 projects a greater number of provinces across the AOR to have modest conflict risk than the other scenarios; that is, there are fewer provinces with very high or very low projected risk.

SSP 5 is the only scenario that is compatible with both RCP scenarios and is thus the best way to compare how changing assumptions about emissions and greenhouse gas concentrations affect the model's conflict forecasts. Figure 2.7 displays projected conflict risk statistics across the AOR by year to facilitate comparison between climate scenarios. As illustrated in Table 2.2, conflict projections largely track between the two climate scenarios. This is true in terms of the average risk of conflict that the models assign across the AOR for the period and in terms of the low- and high-end projections. The model does show divergence between the two scenarios roughly between 2040 and 2060, especially at the high end, with the 90th percentile of conflict predictions as much as two percentage points higher in RCP 8.5 as compared with RCP 6.0 for certain years. This is suggestive evidence that the more drastic climate changes that RCP 8.5 foresees could increase conflict risk in the AOR.

Figure 2.8 shows how the differences in risk are distributed across the AOR for these two scenarios. Iran stands out as having a higher projected risk of conflict in RCP 8.5 across all three decades. Projections under RCP 8.5 trend slightly lower for most of the AOR than under RCP 6.0 in the final decade.

This chapter presented our conflict forecasts at the provincial level across the CENTCOM AOR. The forecasts encompass approximately a 40-year period (2030–2070), although we sometimes elect to depict the results by decade. The forecasts show that, under all socioeconomic and climate conditions included in the model, the AOR will be beset by substantial conflict in the next half century. Consistent with other research, the forecasts also show that while there is suggestive evidence that worse climate outcomes will correlate with a greater incidence of conflict between 2040 and 2060, temperature increases and declines in precipitation are not the major drivers of the security environment, according to our model.

That said, there are good reasons to believe existing research and our own conflict forecasts might be undercounting the impact of climate change on conflict. Chapter 3 explains why this could be the case and presents an excursion on our conflict forecasting to illustrate one source of potential undercounts.

Figure 2.6. Average Forecast Conflict Risk Across Representative Concentration Pathway–Shared Socioeconomic Pathway Scenarios by Decade, 2040–2070

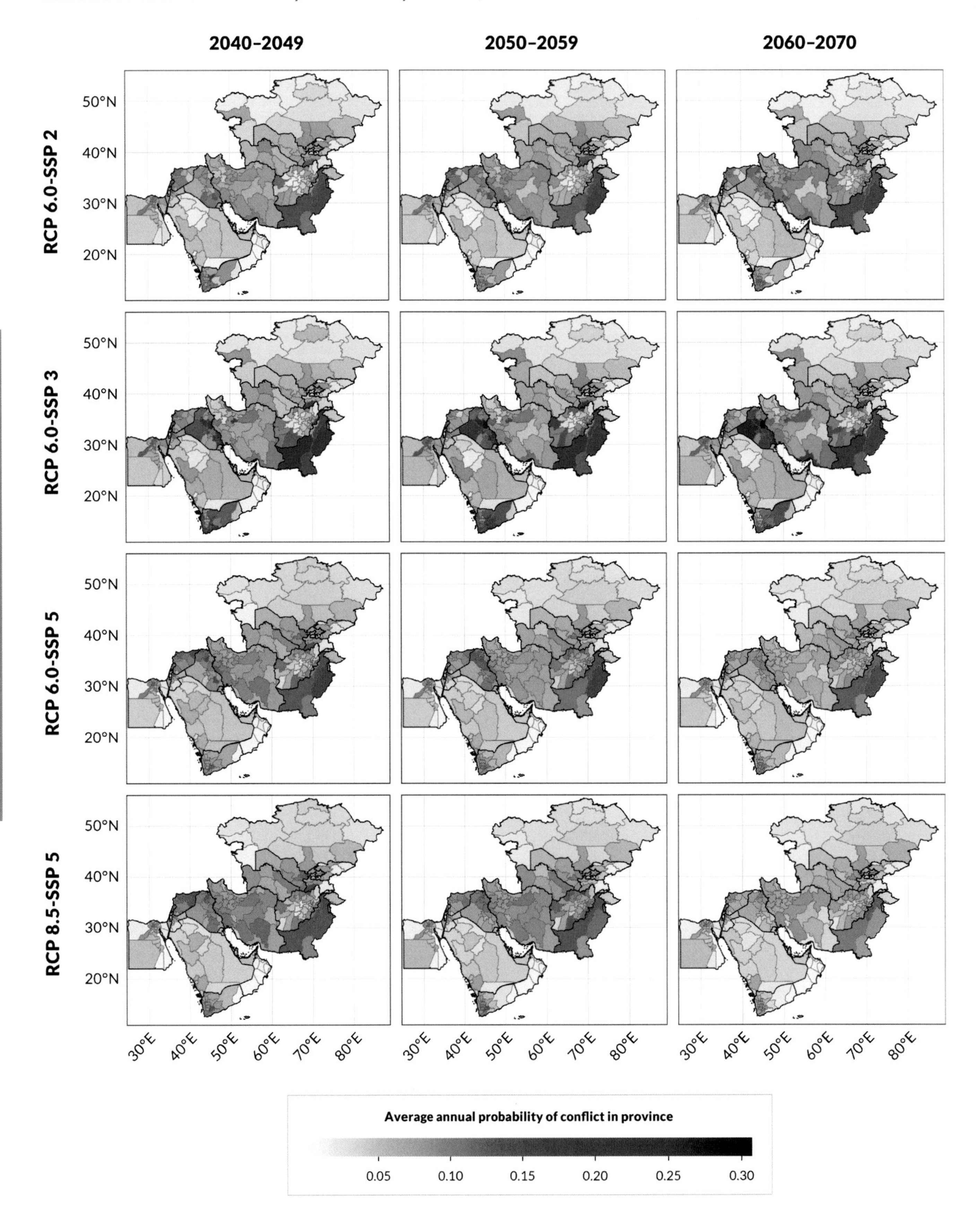

Figure 2.7. Range of Projected Conflict Risk by Year for RCP 6-SSP 5 and RCP 8.5-SSP 5

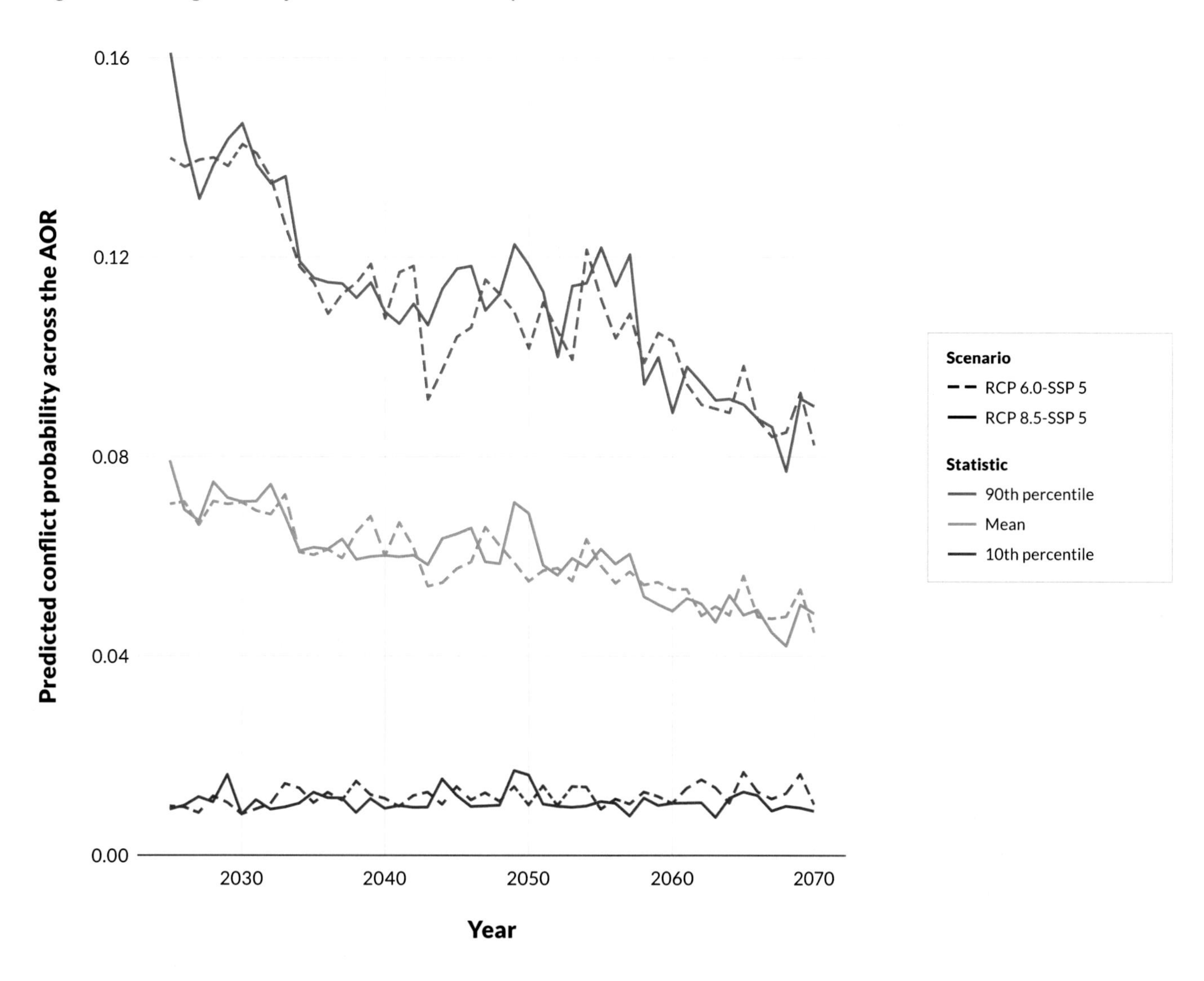

Table 2.2. Projected Conflicts by Decade Under Different Climate Conditions

Scenario	Conflicts AOR-Wide 2040–2049	Conflicts AOR-Wide 2050–2059	Conflicts AOR-Wide 2060–2070
RCP 6.0-SSP 5	16	16	14
RCP 8.5-SSP 5	17	16	13

Figure 2.8. Difference in Projected Conflict Risk, 2040–2049, for RCP 6.0 and RCP 8.5

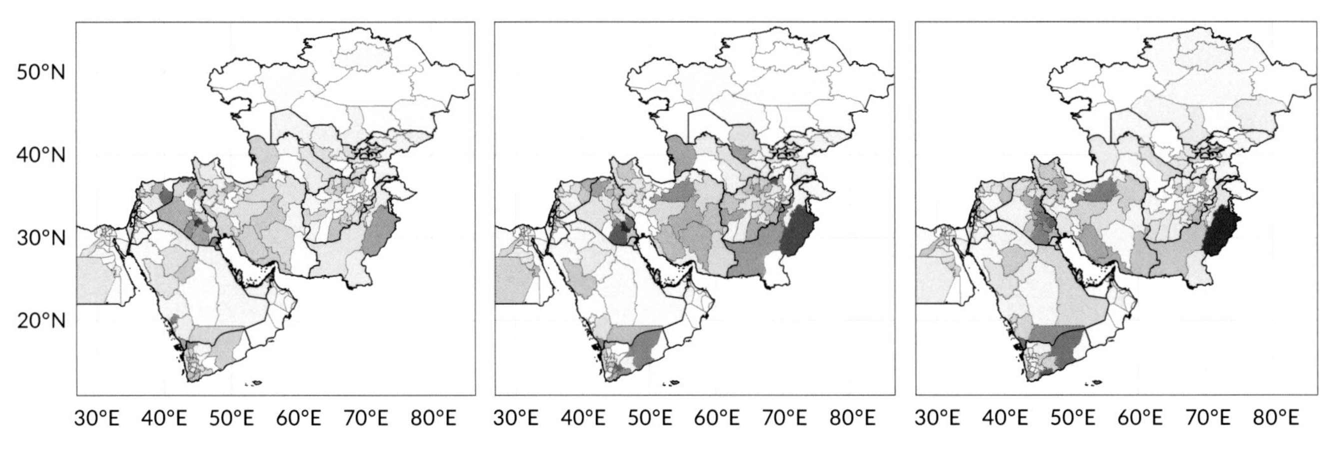

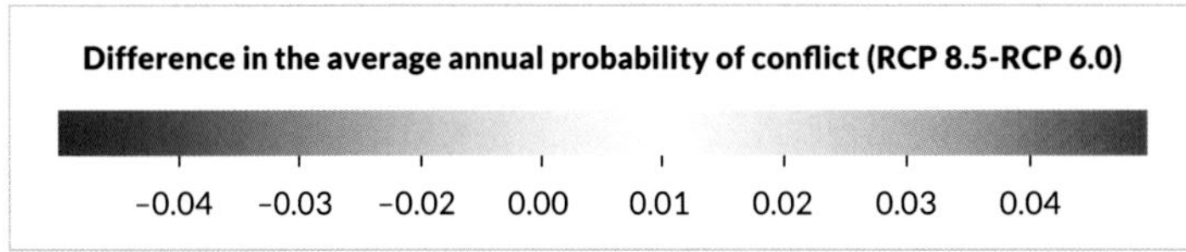

Endnotes

1 Jannis M. Hoch, Sophie P. de Bruin, Halvard Buhaug, Nina Von Uexkull, Rens van Beek, and Niko Wanders, "Projecting Armed Conflict Risk in Africa Towards 2050 Along the SSP-RCP Scenarios: A Machine Learning Approach," *Environmental Research Letters*, Vol. 16, No. 12, December 2021.

2 Witmer et al., 2017.

3 Hegre et al., 2016.

4 Andrew M. Linke, Frank D. W. Witmer, and John O'Loughlin, "Weather Variability and Conflict Forecasts: Dynamic Human-Environment Interactions in Kenya," *Political Geography*, Vol. 92, January 2022.

5 Mariaflavia Harari and Eliana La Ferrara, "Conflict, Climate, and Cells: A Disaggregated Analysis," *Review of Economics and Statistics*, Vol. 100, No. 4, October 2018.

6 Sophie P. de Bruin, Jannis M. Hoch, Nina von Uexkull, Halvard Buhaug, Jolle Demmers, Hans Visser, and Niko Wanders, "Projecting Long-Term Armed Conflict Risk: An Underappreciated Field of Inquiry?" *Global Environmental Change*, Vol. 72, January 2022.

7 Jannis M. Hoch, Sophie de Bruin, and Niko Wanders, "CoPro: A Data-Driven Modelling Framework for Conflict Risk Projections," *Journal of Open Source Software*, Vol. 6, No. 58, February 2021, p. 2855. For an application to conflict in Africa, see Hoch et al., 2021.

8 Ralph Sundberg and Erik Melander, "Introducing the UCDP Georeferenced Event Dataset," *Journal of Peace Research*, Vol. 50, No. 4, July 2013.

9 Stina Högbladh, *UCDP Georeferenced Event Dataset Codebook Version 20.1*, Department of Peace and Conflict Research, Uppsala University, 2020.

10 Högbladh, 2020, pp. 28–30.

11 For more details on CoPro's structure and workflow, see Hoch, de Bruin, and Wanders, 2021, pp. 12–16.

12 For applications of the random forest algorithm to conflict prediction, see David Muchlinski, David Siroky, Jingrui He, and Matthew Kocher, "Comparing Random Forest with Logistic Regression for Predicting Class-Imbalanced Civil War Onset Data," *Political Analysis*, Vol. 24, No. 1, Winter 2016; and Mark Toukan, "International Politics by Other Means: External Sources of Civil War," *Journal of Peace Research*, Vol. 56, No. 6, November 2019.

13 We use administrative provinces as defined by Global Administrative Areas. See Global Administrative Areas, "GADM Maps and Data," database, undated.

14 At the time of our data processing, CMIP6 climate variables were available in the ISIMIP database, but environmental systems models (e.g., land surface and hydrologic models) that use CMIP6 inputs were not. For consistent comparison between climate models and environmental systems models, we used CMIP5 (Coupled Model Intercomparison Project, "CMIP Phase 5 (CMIP5)," database, undated).

15 We selected these indicators on the basis of the availability of historical data that covered the entire reference period. The inclusion of other indicators (for example, indices for aridity or drought) could potentially change the findings, although existing work does not suggest that any particular climatic indicator is reliably predictive of conflict.

16 ISIMIP, "The Inter-Sectoral Impact Model Intercomparison Project," database, undated.

17 For more information on the RCPs, see Detlef P. van Vuuren, Jae Edmonds, Mikiko Kainuma, Keywan Riahi, Allison Thomson, Kathy Hibbard, George C. Hurtt, Tom Kram, Volker Krey, Jean-Francois Lamarque, Toshihiko Masui, Malte Meinshausen, Nebojsa

Nakicenovic, Steven J. Smith, and Steven K. Rose, "The Representative Concentration Pathways: An Overview," *Climatic Change*, Vol. 109, August 2011.

18 IPCC, 2022, p. 137.

19 Michelle E. Miro, Flannery Dolan, Karen M. Sudkamp, Jeffrey Martini, Karishma V. Patel, and Carlos Calvo Hernandez, *A Hotter and Drier Future Ahead: An Assessment of Climate Change in U.S. Central Command*, RAND Corporation, RR-A2338-1, 2023.

20 Halvard Buhaug and Kristian Skrede Gleditsch, "Contagion or Confusion? Why Conflicts Cluster in Space," *International Studies Quarterly*, Vol. 52, No. 2, June 2008.

21 Detlef P. van Vuuren, Elmar Kriegler, Brian C. O'Neill, Kristie L. Ebi, Keywan Riahi, Timothy R. Carter, Jae Edmonds, Stephane Hallegatte, Tom Kram, Ritu Mathur, and Harald Winkler "A New Scenario Framework for Climate Change Research: Scenario Matrix Architecture," *Climatic Change*, Vol. 122, February 2014.

22 J. Gao, "Global 1-km Downscaled Population Base Year and Projection Grids Based on the Shared Socioeconomic Pathways, v1.01 (2000–2100)," database, Socioeconomic Data and Applications Center, National Aeronautics and Space Administration, 2020; Daisuke Murakami, Takahiro Yoshida, and Yoshiki Yamagata, "Gridded GDP Projections Compatible with the Five SSPs (Shared Socioeconomic Pathways)," *Frontiers in Built Environment*, Vol. 7, October 2021, p. 138; Marina Andrijevic, Jesus Crespo Cuaresma, Raya Muttarak, and Carl-Friedrich Schleussner, "Governance in Socioeconomic Pathways and Its Role for Future Adaptive Capacity," *Nature Sustainability*, Vol. 3, No. 1, January 2020. They produce model-based projections of "a composite governance index based on six categories: voice and accountability, political stability, government effectiveness, regulatory quality, rule of law and control of corruption" (Andrijevic et al., 2020, p. 40).

Note that some of the administrative provinces used have a smaller area. In these instances, the values of the conflict, climate, and socioeconomic indicators sampled at the resolution of a half-degree grid cell will be applied to more than one province. This could influence modeling results, particularly those that concern the persistence of conflict across time. As Figure 2.4 shows, prior year conflict in neighboring provinces is a more important feature than prior year conflict in the same province. This result could be an outcome of some units of observation being at a finer resolution than the sampled indicators.

23 van Vuuren et al., 2014, p. 382.

24 Christopher Blattman and Edward Miguel, "Civil War," *Journal of Economic Literature*, Vol. 48, No. 1, March 2010; Håvard Hegre and Nicholas Sambanis, "Sensitivity Analysis of Empirical Results on Civil War Onset," *Journal of Conflict Resolution*, Vol. 50, No. 4, August 2006.

25 Hoch et al., 2021, p. 17.

26 The area under the curve of the ROC-AUC score indicates how well the model discriminates between conflict and non-conflict cases. It reflects the chances that the model will assign a higher probability of conflict to a random case of conflict than to a non-case of conflict.

27 For ease of visualization, Figure 2.6 displays the final three decades of the forecast.

28 Brian C. O'Neill, Elmar Kriegler, Keywan Riahi, Kristie L. Ebi, Stephane Hallegatte, Timothy R. Carter, Ritu Mathur, and Detlef P. van Vuuren, "A New Scenario Framework for Climate Change Research: The Concept of Shared Socioeconomic Pathways," *Climatic Change*, Vol. 122, No. 3, February 2014.

29 Zeke Hausfather and Glen P. Peters, "Emissions—The 'Business as Usual' Story Is Misleading," *Nature*, January 29, 2020.

30 O'Neill et al., 2014, p. 398.

CHAPTER 3

UNCERTAINTY OVER CLIMATE RISK

MUCH OF THE EXISTING scholarly literature on the link between climate change and violent conflict employs state-of-the-art statistical techniques, highly detailed data, and clever research designs. This field, however, is an inherently challenging one to study. This chapter reviews four challenges that scholars and other analysts have faced. While progress has been made in addressing these challenges, considerable uncertainty remains. From the perspective of military decisionmakers and planners, this uncertainty might obscure the risks that climate change will contribute to conflict at a level beyond what was anticipated by many current analyses. We conclude this chapter with an analytic excursion on our own modeling (presented in Chapter 2) that illustrates this risk.

Challenges to Accurately Forecasting Climate-Related Conflict

The scholarly literature on the climate-conflict link has followed a trajectory that is typical of such research agendas. Analysis began with qualitative inquiries into the potential links between climate change and violent conflict.[1] When these early analyses suggested that there were several plausible links, scholars began to develop quantitative models to test how frequently these links appeared. The initial models tended to test very simple and easy to measure relationships, such as whether particularly high or low average temperatures or rainfall in one period and in one place were associated with increased levels of violence in the same place in the subsequent period. Such analyses failed to yield consistent findings about the hypothesized link between climate and conflict; some analyses found increases in violence, some found no relationship, and some even found declines. More consistent findings, however, indicated that such factors as the level of economic development or the quality of governance were much more important drivers of peace and conflict than climate change.

There were at least four potential problems with these models:

- The link between climate and conflict was neither unidirectional nor direct; rather, the presence of conflict limits a state's ability to adapt to climate change, further increasing the risk of conflict. Additionally, climate hazards could suppress economic development, contributing to conflict via economic conditions.[2]
- The models required historical data to produce results. Climate change, however, might contribute to future conditions that are fundamentally different from those that were characterized by the recent past.[3]
- The models frequently looked for changes in levels of violence in the same geographic unit (e.g., a country, an administrative subdivision, or a grid-square of a map) in which an adverse climate event had occurred. But such processes as climate-induced migration might contribute to violence in places that are far removed from the initial climate event.
- The models frequently examined the relationship between an adverse climate event in one period and conflict levels in the period immediately following. Many of the pathways linking climate factors to violence, however, are complex and only unfold over longer periods, often many years.[4]

Over time, social scientists have refined these models, developing ever more nuanced measures and sophisticated ways of modeling the complex paths that link climate and conflict. None of these challenges, however, have been fully resolved. The remainder of this chapter analyzes whether these persistent challenges might obscure risks that U.S. defense planners should incorporate into their plans.

Data and Measurement

Many early statistical analyses of the relationship between climate and conflict relied on fairly blunt measures of adverse climate-related events, such as particularly high or low temperatures or levels of precipitation in any given month or year. The results of these early studies were generally highly contradictory; different studies came to different conclusions. The reasons for this lack of consensus might reflect in part the bluntness of these measures.[5]

More recent work has begun to disaggregate these measures. For example, such events as droughts or excessive rain are not likely to have the same effects throughout the year; they could have devastating consequences at critical points in the growing season and much milder effects at other times of year. Recent studies have examined more sophisticated versions of the climate-conflict link, such as the effects of seasonality across multiple crops in multiple countries and for specific crops in specific countries. These studies have tended to find stronger relationships between conflict and climate.[6]

In addition to the challenges of data measurement around climate variables, there are corresponding challenges in relating conflict outcomes. Because it is so difficult to obtain accurate counts of the number of people killed in armed conflicts, analyses of conflict intensity are much less frequent than analyses of conflict onset. But climate hazards might intensify existing conflicts rather than (or in addition to) generating new conflicts.[7] Consequently, statistical analyses might understudy one of the more prevalent links between climate change and violence.

Focusing on Extreme Events

One way to test whether future climate change could lead to a different relationship with conflict is to focus on extreme weather events. These historic outliers are projected to become more frequent in the coming decades, with so-called once in a century storms becoming once in a decade storms and culminating in ever more devasting consequences.

Some analysts have implemented research designs that help isolate the apparent consequences of more extreme events. One analysis of extremely hot temperatures found that they were not associated with violence *overall* during the past three decades, but they *were* associated with increased violence when they occurred at the hottest times of the year and in the hottest regions of the Middle East and Africa.[8] Other analyses exploited the highly detailed data available on floods to better assess the impact of extreme events. Most of those studies found that particularly severe floods were associated with higher levels of political unrest and with the prolongation of existing wars.[9] Other research finds that transnational terrorism increases in the wake of natural disasters.[10] These studies thus suggest that a future characterized by what were once extreme climate-related events will become at least somewhat more prone to conflict.

The Spatial Diffusion of Conflict Potential

Climate-related disasters might contribute to conflict in areas that are distant from the initial event, including across borders. Adverse climate events (such as droughts or floods) might contribute to distant conflicts through either of two mechanisms: (1) climate change-induced migration and (2) climate change-induced food price shocks.[11] While early statistical studies in the field largely failed to account for non-proximate geographical effects, modeling techniques have grown increasingly sophisticated in this regard. Broader studies of armed conflict have generally found support for the proposition that conflicts in one area can increase the likelihood of conflict in neighboring areas.[12]

Therefore, studies that solely examine changes in levels of violence in the same geographic area that experienced an adverse climate event might not capture the full scope of the climate-conflict relationship.

Complex, Long-Running Processes of Destabilization

Some forms of violence, such as violent crime or riots, can occur spontaneously. Larger-scale, organized violence, such as insurgencies or civil wars, typically takes longer to evolve. Communities must become disaffected with the current political order, become aware of others' discontent and readiness for action, organize fighting units, procure the arms necessary to fight, and develop sustainment networks to support the fighting. At the same time, the security forces that support the current government might need to weaken to the point that they can no longer suppress internal violence. If none of these preconditions is in place, it might take many years between when a climate-related disaster provides an incentive and opening for violence and when large-scale violence emerges.

Statistical models that look for a relationship between an adverse climate event in one period (such as a year or a month) and violence in the subsequent period would miss these more slowly evolving dynamics and might therefore understate the future risk posed by climate change-related shocks and stressors. Some statistical analyses that examine longer periods have found evidence that such events as floods can lead to heightened levels of violence for several years afterward.[13] These long-running processes that might connect climate-related events to armed conflict could operate through political delegitimization, the breakdown of intercommunal relations, and economic losses.[14]

An Excursion: Modeling the Effect of Drought on Conflict

On the basis of the preceding analysis of why existing studies—including our own—might be underestimating the strength of the climate-conflict link, we return to our machine learning model to interrogate one of the model's underlying assumptions. The SSPs that are used widely in the field of climate change (and in our own machine learning model) project future growth rates on the basis of historical averages. These historical averages, however, smooth out the ups and downs of business cycles and economic shocks, portraying future trends as an uninterrupted progression of economic growth. Because economic development and growth are among the most important predictors of conflict, using such smoothed averages omits a key source of instability and thus leads to unrealistically optimistic estimates of future conflict.[15]

Addressing this shortcoming in forecasting techniques is challenging because it is impossible to predict what specific factors might trigger future economic downturns or when those downturns are likely to occur. Rather than attempting to systematically introduce all possible causes of economic downturns into our model, we instead offer a single illustration that is based on one of the most intuitive pathways linking climate change to economic decline: severe drought in agriculturally dependent regions. Drought might increase the risk of conflict by acting as a negative income shock in certain provinces and lead those who are negatively affected to turn to violence to address their grievance or lead to waves of migration, which in turn can produce friction and competition that could lead to conflict risk.

For this vignette, we selected RCP 6.0-SSP 3 as the baseline case. To model the effect of drought on GDP, we penalized the projected GDP growth of grid cells during drought years. We defined drought years as those country-years that experienced a greater than two standard deviation shortfall in precipitation on a rolling ten-year mean. Growth is reduced in each grid cell to achieve an overall country-level GDP growth rate reduction of 1.5 percentage points in drought years. We derived the 1.5 percent figure from a characterization of the existing literature on drought's effect on economic growth.[16] Lower or higher figures are plausible, especially when considering the variation across and within countries along dimensions that would make areas vulnerable to drought impacts. The goal of this exercise was to investigate a directional effect from a climatic effect to conflict, and the correlation—via economic growth to conflict risk—is well demonstrated in existing work. While greater (lesser) growth penalties would likely result in higher (lower) conflict risks, we adopted the 1.5 percentage point estimate for demonstrative purposes.

Cells take a penalty to GDP growth that is proportional to their share of the country's overall agricultural GDP output, so more agriculturally dependent cells bear a greater cost.[17] In the wake of drought years, the GDP growth path is kept in line with assumptions under SSP 3.[18] Overall GDP levels are lower toward the end of the projected period for those countries that are forecast to experience drought, as negative growth shocks add up over time. We also recomputed the forecast governance quality under SSP 3 because the model uses overall country GDP as an input to the index of governance quality.[19]

Figure 3.1 shows the aggregate difference in conflict across the AOR when comparing the unadjusted projections of our model and projections that account for the impact of drought. The figure reveals a consistently greater risk when the growth reducing effects of drought are modeled. Figure 3.1 suggests that the expected number of conflicts is higher for each decade between 2040 and 2070. Figure 3.2 displays the difference in projected conflict risk by decade and by province and reveals that the difference is significantly driven by large increases in Iraq, Yemen, and Pakistan, although modest increases in risk are projected for other countries and provinces as well.

This vignette illustrates one way in which the existing SSP frameworks and associated forecasts of socioeconomic indicators could understate future risks of conflict. It is important to note that this vignette does not model every pathway from drought to conflict, nor does it model the full variety of effects that precipitation and other climate changes could have on income, governance, or population levels or distribution. The vignette does, however, indicate that climate changes under plausible scenarios could result in future conflict risks that are substantially worse than the baseline scenarios discussed above.

Figure 3.1. Difference in Area of Responsibility Conflict Risk When Accounting for the Effects of Drought on Gross Domestic Product

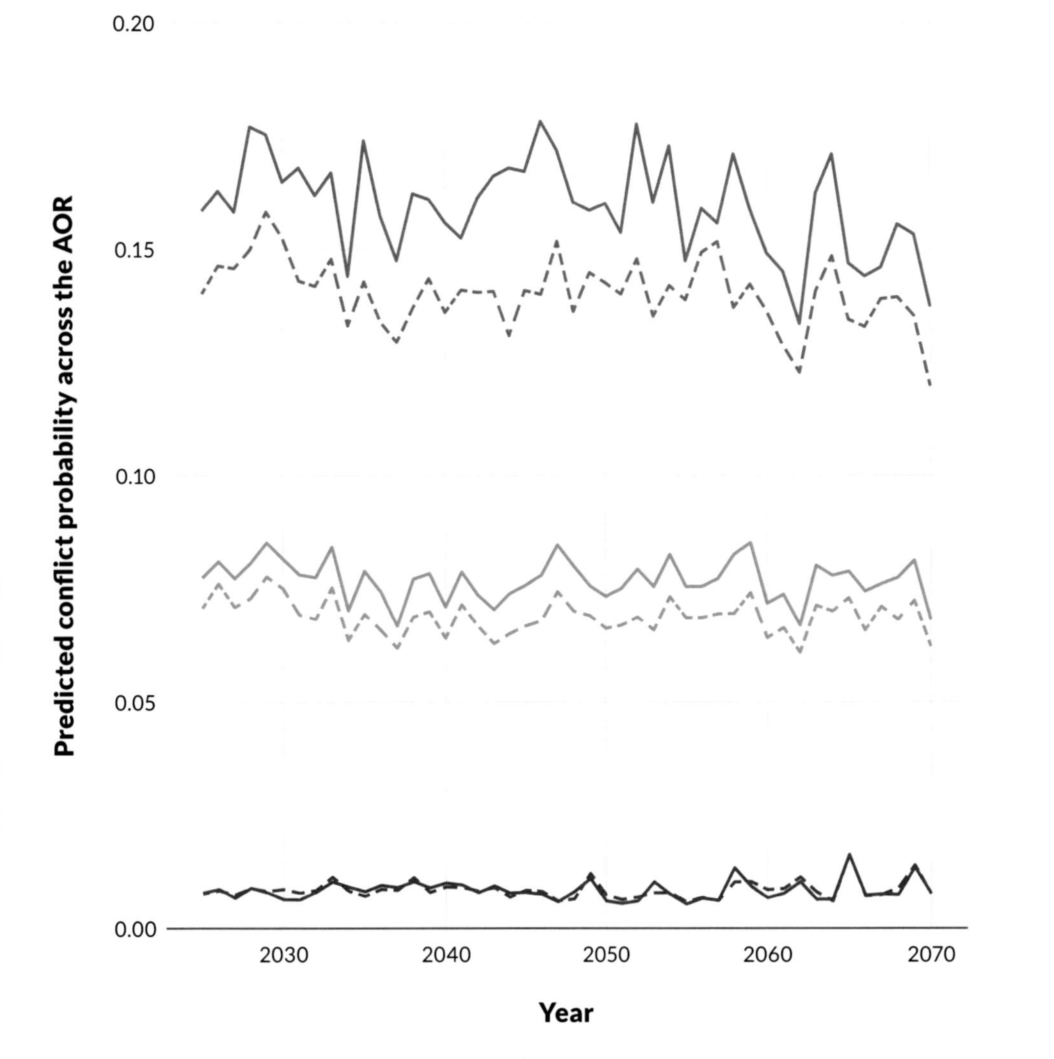

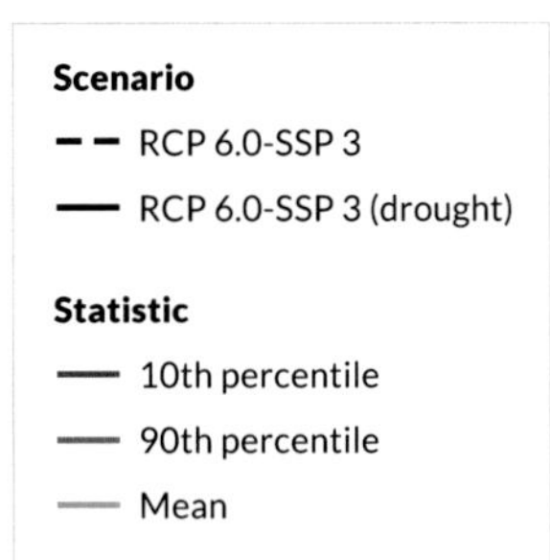

This potential for undercounting conflict is made worse by the "feedback loop" between conflict and economic development. As noted above, GDP is both a predictor of conflict and affected by conflict. Researchers have estimates that use plausible forecasts of conflict at the country level that the SSPs may overestimate GDP per capita growth by between 25 and 30 percent by not accounting for the feedback loop between conflict and economic growth.[20] Notably, this problem is worse for countries with a history of conflict, as is the case for many countries in the CENTCOM area.[21] The results in this chapter do not employ a dynamic model of the climate-growth relationship. A more complicated model that accounted for this dynamic would likely result in even higher forecasts of conflict risk across the AOR.

Conclusion

This chapter does not invalidate the consensus on the climate-conflict link captured in the IPCC and U.S. IC documents reviewed in Chapter 1. Those documents uniformly stressed that there is likely a relationship between climate change and conflict, although the extent of that relationship is highly uncertain. This chapter has highlighted several reasons why there is at least the potential for the levels of climate-related conflict to be significantly worse than is commonly believed. Military decisionmakers and planners will need to account for such uncertainty in their planning.

Figure 3.2. Difference in Province-Level Conflict Risk When Accounting for the Effects of Drought on Gross Domestic Product

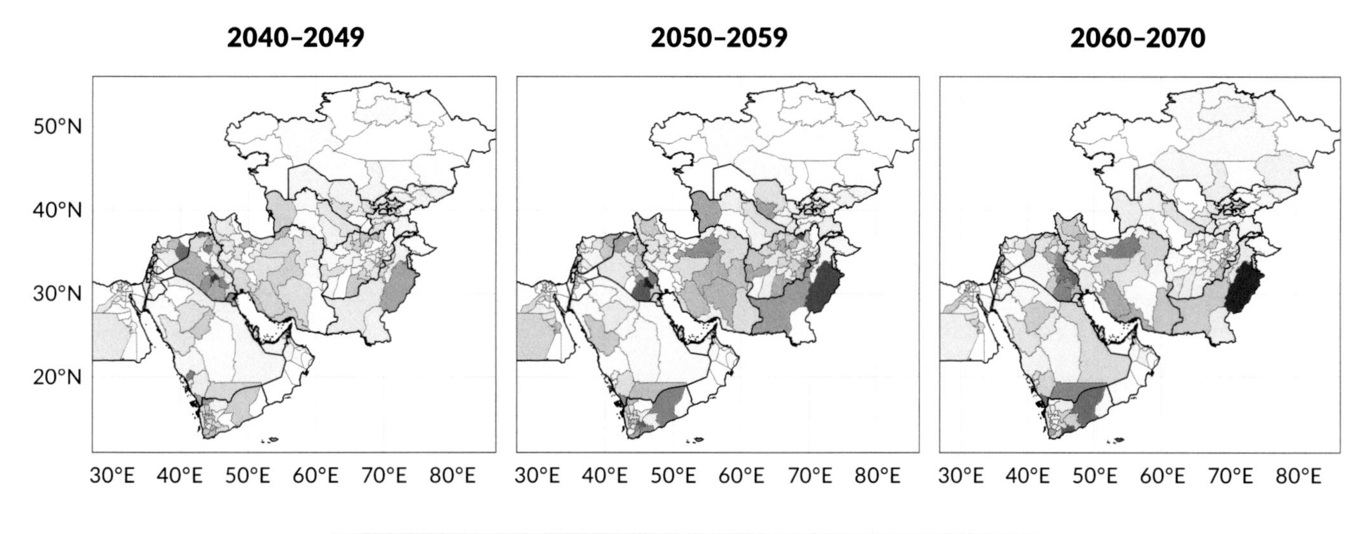

Endnotes

1 See Thomas F. Homer-Dixon, *Environment, Scarcity, and Violence*, Princeton University Press, 1999; and Colin H. Kahl, *States, Scarcity, and Civil Strife in the Developing World*, Princeton University Press, 2006.

2 Halvard Buhaug, "Climate-Conflict Research: Some Reflections on the Way Forward," *WIREs Climate Change*, Vol. 6, No. 3, May/June 2015; Halvard Buhaug and Nina von Uexkull, "Vicious Circles: Violence, Vulnerability, and Climate Change," *Annual Review of Environment and Resources*, Vol. 46, October 2021.

3 Buhaug, 2015, p. 270.

4 Buhaug, 2015, p. 273.

5 See Buhaug, 2015, p. 270.

6 Elizabeth A. Mack, Erin Bunting, James Herndon, Richard A. Marcantonio, Amanda Ross, and Andrew Zimmer, "Conflict and its Relationship to Climate Variability in Sub-Saharan Africa," *Science of the Total Environment*, Vol. 775, June 25, 2021; Raul Caruso, Ilaria Petrarca, and Roberto Ricciuti, "Climate Change, Rice Crops, and Violence: Evidence from Indonesia," *Journal of Peace Research*, Vol. 53, No. 1, January 2016.

7 For a recent study that looks at conflict intensity (vice onset), see Tobias Ide, "Rise or Recede? How Climate Disasters Affect Armed Conflict Intensity," *International Security*, Vol. 47, No. 4, Spring 2023.

8 David Helman and Benjamin F. Zaitchik, "Temperature Anomalies Affect Violent Conflicts in African and Middle Eastern Warm Regions," *Global Environmental Change*, Vol. 63, July 2020.

9 Tobias Ide, Anders Kristensen, and Henrikas Bartusevičius, "First Comes the River, Then Comes the Conflict? A Qualitative Comparative Analysis of Flood-Related Political Unrest," *Journal of Peace Research*, Vol. 58, No. 1, January 2021; Ramesh Ghimire and Susana Ferreira, "Floods and Armed Conflict," *Environment and Development Economics*, Vol. 21, No. 1, February 2016; Ramesh Ghimire, Susana Ferreira, and Jeffrey H. Dorfman, "Flood-Induced Displacement and Civil Conflict," *World Development*, Vol. 66, February 2015. One of the studies reviewed did not find an association between disaster intensity and conflict (Drago Bergholt and Päivi

Lujala, "Climate-Related Natural Disasters, Economic Growth, and Armed Civil Conflict," *Journal of Peace Research*, Vol. 49, No. 1, January 2012).

10 Jomon Aliyas Paul and Aniruddha Bagchi, "Does Terrorism Increase After a Natural Disaster? An Analysis Based upon Property Damage," *Defence and Peace Economics*, Vol. 29, No. 4, July 2018.

11 Rafael Reuveny, "Climate Change-Induced Migration and Violent Conflict," *Political Geography*, Vol. 26, No. 6, August 2007; Tor A. Benjaminsen, "Does Supply-Induced Scarcity Drive Violent Conflicts in the African Sahel? The Case of the Tuareg Rebellion in Northern Mali," *Journal of Peace Research*, Vol. 45, No. 6, November 2008; Cullen S. Hendrix and Stephan Haggard, "Global Food Prices, Regime Type, and Urban Unrest in the Developing World," *Journal of Peace Research*, Vol. 52, No. 2, March 2015; Charles P. Martin-Shields and Wolfgang Stojetz, "Food Security and Conflict: Empirical Challenges and Future Opportunities for Research and Policy Making on Food Security and Conflict," *World Development*, Volume 119, July 2019; Clionadh Raleigh, Hyun Jin Choi, and Dominic Kniveton, "The Devil Is in the Details: An Investigation of the Relationships Between Conflict, Food Price and Climate Across Africa," *Global Environmental Change*, Vol. 32, May 2015.

12 Hanne Fjelde, "Farming or Fighting? Agricultural Price Shocks and Civil War in Africa," *World Development*, Vol. 67, March 2015.

13 Ghimire, Ferreira, and Dorfman, 2015.

14 Ryan E. Carlin, Gregory J. Love, and Elizabeth J. Zechmeister, "Natural Disaster and Democratic Legitimacy: The Public Opinion Consequences of Chile's 2010 Earthquake and Tsunami," *Political Research Quarterly*, Vol. 67, No. 1, March 2014; Nina von Uexkull, Marco d'Errico, and Julius Jackson, "Drought, Resilience, and Support for Violence: Household Survey Evidence from DR Congo," *Journal of Conflict Resolution*, Vol. 64, No. 10, November 2020; Carl-Friedrich Schleussner, Jonathan F. Donges, Reik V. Donner, and Hans Joachim Schellnhuber, "Armed-Conflict Risks Enhanced by Climate-Related Disasters in Ethnically Fractionalized Countries," *Proceedings of the National Academy of Sciences*, Vol. 113, No. 33, July 2016; Buhaug and von Uexkull, 2021; Pieter Serneels and Marijke Verpoorten, "The Impact of Armed Conflict on Economic Performance: Evidence from Rwanda," *Journal of Conflict Resolution*, Vol. 59, No. 4, June 2015.

15 Halvard Buhaug and Jonas Vestby, "On Growth Projections in the Shared Socioeconomic Pathways," *Global Environmental Politics*, Vol. 19, No. 4, November 2019.

16 It is well-supported that droughts have an overall negative effect on GDP and GDP growth and that the impact is stronger for the agricultural sector, as droughts can lead to crop death, migration away from agricultural areas, and declines in livestock productivity. We estimated the damage of a moderate to severe drought to have a negative impact on the overall GDP of a country that ranges between 0.2 percent and 4.4 percent, with clustering around a 1 percent impact on GDP and an average of 1.48 percent. See Thomas Fomby, Yuki Ikeda, and Norman Loayza, "The Growth Aftermath of Natural Disasters," World Bank, Policy Research Working Paper No. 5002, July 2009; Claudio Raddatz, "The Wrath of God: Macroeconomic Costs of Natural Disasters," World Bank, Policy Research Working Paper 5039, September 2009; Habibollah Salami, Naser Shahnooshi, and Kenneth J. Thomson, "The Economic Impacts of Drought on the Economy of Iran: An Integration of Linear Programming and Macroeconometric Modelling Approaches," *Ecological Economics*, Vol. 68, No. 4, February 2009; Perrihan Al-Riffai, Clemens Breisinger, Dorte Verner, and Tingju Zhu, "Droughts in Syria: An Assessment of Impacts and Options for Improving the Resilience of the Poor," *Quarterly Journal of International Agriculture*, Vol. 51, No. 1, February 2012.

17 We used the World Bank's gridded estimates of agricultural GDP to estimate the total agricultural GDP of grid cells as of 2010 (Yating Ru, Brian Blankespoor, Ulrike Wood-Sichra, Timothy S. Thomas, Liangzhi You, and Erwin Kalvelagen, "Estimating Local Agricultural Gross Domestic Product [AgGDP] Across the World," *Earth System Science Data*, Vol. 15, No. 3, March 2023).

18 This was achieved by calculating the year-on-year growth of GDP under SSP 3 assumptions and calculating GDP forecasts for each cell on the basis of its initial forecast value and the adjusted growth rate after accounting for drought impacts.

19 We reran the model to produce the governance indicators using all the original data sources and made appropriate changes to country-year GDP estimates according to the projected drought years in our data. The repository to compute the governance indicators is available at Marina Andrijevic, "governance2019," database, 2019, updated October 23, 2022.

20 Kristina Petrova, Gudlaug Olafsdottir, Håvard Hegre, and Elisabeth A. Gilmore, "The 'Conflict Trap' Reduces Economic Growth in the Shared Socioeconomic Pathways," *Environmental Research Letters*, Vol. 18, No. 2, January 2023.

21 Petrova et al., 2023, p. 8.

CHAPTER 4

CONCLUSION

THIS REPORT IS INTENDED to help CENTCOM leadership and planners and their interagency partners better understand plausible patterns of conflict in the CENTCOM AOR during the 2030–2070 period. Because climate hazards will intensify over the same period, our forecasts integrate several climate variables (i.e., temperature and precipitation) into the outcomes. To account for uncertainty in both climate and socioeconomic projections, we use ranged forecasts that vary underlying assumptions about both sets of variables (i.e., climate and non-climate).

We find that, under all the socioeconomic and climate conditions that we considered, the CENTCOM AOR will experience substantial conflict in the next half century. While there is suggestive evidence that worse climate outcomes will correlate with a greater incidence of conflict between 2040 and 2060, temperature increases and declines in precipitation are not the major drivers of the future security environment according to our machine learning model. Rather, where climate hazards increase conflict risk, they do so by interacting with other variables that are stronger predictors of conflict. The strongest predictors of conflict in our model are (1) governance, although the governance assumptions are partly based on economic performance, and (2) the presence of conflict in a neighboring area the prior year. Economic performance is the third most important factor, although again, its effect is partly reflected through the impact of governance on conflict outcomes.

That said, there are good reasons to believe that existing research and our own conflict forecasts might be underestimating the impact of climate variables on conflict. The main limitations of existing research are inadequate attention given to the dynamic relationship between climate hazards, the economy, and conflict that could result in negative feedback loops. Specifically, the presence of conflict limits a state's ability to adapt to climate change, further increasing its risk of conflict traps. Additionally, climate hazards could suppress economic development, contributing to conflict via socioeconomic conditions. Climate change could also contribute to conditions that shape conflict risk in a manner that is fundamentally different from conditions that characterized the recent past. Finally, climate hazards could—via migration or food price shocks—generate conflict that is far from the localized climate impacts or could result in conflict in future periods that would not be captured in some existing research.

To test whether our own modeling might be underestimating the strength of the climate-conflict relationship, we undertook an additional modeling excursion that factors in the economic impact of drought. After making assumptions that are grounded in existing research about the impact of drought on the economies of agriculture-dependent areas, we project a significant increased risk of conflict in those areas. This excursion suggests that our baseline forecasts might be underestimating conflict risk that is attributed to climate change.

One limitation of our conflict modeling is that it does not measure the *intensity* of conflict beyond the threshold of a minimum of 25 annual battle deaths—in line with much existing research—for violence to qualify as a conflict. Therefore, our forecasts cannot be used to test the proposition that climate hazards could exacerbate existing conflicts rather than generate new ones. We also could not model interstate war because that remains a rare phenomenon. Finally, data limitations precluded us from dynamically modeling the impact of conflict on future economic growth, which in turn could drive more conflict.

It is important to note that intrastate conflicts generally arise from local phenomena and because of interactions between contingent events and larger structural factors, including both climate and non-climate factors. Planners must therefore be attentive to factors at the province-level and below that could be indicative of conflict risk. The ways in which climate and non-climate factors interact to shape conflict risk will also vary at the province level and below. Thus, understanding conflict risk at the level of country or region must be accompanied by attentiveness to local dynamics.

The current scholarly consensus might be right that climate is a relatively minor and uncertain predictor of future conflict. The research in this report, however, suggests that there is significant risk that the future effects of climate could be worse than the consensus implies. Although they should not overstate the dangers that climate change poses to U.S. national security, CENTCOM leaders and planners would be prudent to take actions to hedge against this risk.

Since the future of conflict will be shaped not only by the variables treated in these forecasts, the next report in this series addresses how U.S. competitors—Iran in the CENTCOM AOR and Russia and China as extra-regional actors—might attempt to exploit climate change to advance their security interests in the region. That report is titled *Mischief, Malevolence, or Indifference? How Adversaries Consider Climate-Induced Conflict in U.S. Central Command.* The fifth and final report in this series focuses on the implications of climate and conflict for CENTCOM, both from the standpoint of what OAIs could be undertaken to mitigate the risk of conflict onset and what OAIs might be necessary to respond to future conflicts that do occur.

ABBREVIATIONS

AOR	area of responsibility
CENTCOM	U.S. Central Command
CMIP6	Coupled Model Intercomparison Project
GDP	gross domestic product
GED	Georeferenced Event Dataset
IC	intelligence community
IPCC	Intergovernmental Panel on Climate Change
ISIMIP	Inter-Sectoral Impact Model Intercomparison Project
OAIs	operations, activities, and investments
RCP	Representative Concentration Pathway
SSP	Shared Socioeconomic Pathway

REFERENCES

Al-Riffai, Perrihan, Clemens Breisinger, Dorte Verner, and Tingju Zhu, "Droughts in Syria: An Assessment of Impacts and Options for Improving the Resilience of the Poor," *Quarterly Journal of International Agriculture*, Vol. 51, No. 1, February 2012.

Andrijevic, Marina, "governance2019," database, 2019, updated October 23, 2022. As of July 31, 2023: https://github.com/marina-andrijevic/governance2019

Andrijevic, Marina, Jesus Crespo Cuaresma, Raya Muttarak, and Carl-Friedrich Schleussner, "Governance in Socioeconomic Pathways and Its Role for Future Adaptive Capacity," *Nature Sustainability*, Vol. 3, No. 1, January 2020.

Benjaminsen, Tor A., "Does Supply-Induced Scarcity Drive Violent Conflicts in the African Sahel? The Case of the Tuareg Rebellion in Northern Mali," *Journal of Peace Research*, Vol. 45, No. 6, November 2008.

Bergholt, Drago, and Päivi Lujala, "Climate-Related Natural Disasters, Economic Growth, and Armed Civil Conflict," *Journal of Peace Research*, Vol. 49, No. 1, January 2012.

Blattman, Christopher, and Edward Miguel, "Civil War," *Journal of Economic Literature*, Vol. 48, No. 1, March 2010.

Buhaug, Halvard, "Climate–Conflict Research: Some Reflections on the Way Forward," *WIREs Climate Change*, Vol. 6, No. 3, May/June 2015.

Buhaug, Halvard, and Kristian Skrede Gleditsch, "Contagion or Confusion? Why Conflicts Cluster in Space," *International Studies Quarterly*, Vol. 52, No. 2, June 2008.

Buhaug, Halvard, and Jonas Vestby, "On Growth Projections in the Shared Socioeconomic Pathways," *Global Environmental Politics*, Vol. 19, No. 4, November 2019.

Buhaug, Halvard, and Nina von Uexkull, "Vicious Circles: Violence, Vulnerability, and Climate Change," *Annual Review of Environment and Resources*, Vol. 46, October 2021.

Carlin, Ryan E., Gregory J. Love, and Elizabeth J. Zechmeister, "Natural Disaster and Democratic Legitimacy: The Public Opinion Consequences of Chile's 2010 Earthquake and Tsunami," *Political Research Quarterly*, Vol. 67, No. 1, March 2014.

Caruso, Raul, Ilaria Petrarca, and Roberto Ricciuti, "Climate Change, Rice Crops, and Violence: Evidence from Indonesia," *Journal of Peace Research*, Vol. 53, No. 1, January 2016.

Chandler, Nathan, Jeffrey Martini, Karen M. Sudkamp, Maggie Habib, Benjamin J. Sacks, and Zohan Tariq, *Pathways from Climate Change to Conflict in U.S. Central Command*, RAND Corporation, RR-A2338-2, 2023.

Coupled Model Intercomparison Project, "CMIP Phase 5 (CMIP5)," database, undated. As of August 8, 2023: https://wcrp-cmip.org/cmip-phase-5-cmip5/

de Bruin, Sophie P., Jannis M. Hoch, Nina von Uexkull, Halvard Buhaug, Jolle Demmers, Hans Visser, and Niko Wanders, "Projecting Long-Term Armed Conflict Risk: An Underappreciated Field of Inquiry?" *Global Environmental Change*, Vol. 72, January 2022.

Fjelde, Hanne, "Farming or Fighting? Agricultural Price Shocks and Civil War in Africa," *World Development*, Vol. 67, March 2015.

Fomby, Thomas, Yuki Ikeda, and Norman Loayza, "The Growth Aftermath of Natural Disasters," World Bank, Policy Research Working Paper No. 5002, July 2009.

Gao, J., "Global 1-km Downscaled Population Base Year and Projection Grids Based on the Shared Socioeconomic Pathways, v1.01 (2000–2100)," database, Socioeconomic Data and Applications Center, National Aeronautics and Space Administration, 2020. As of May 22, 2023: https://sedac.ciesin.columbia.edu/data/set/popdynamics-1-km-downscaled-pop-base-year-projection-ssp-2000-2100-rev01

Ghimire, Ramesh, and Susana Ferreira, "Floods and Armed Conflict," *Environment and Development Economics*, Vol. 21, No. 1, February 2016.

Ghimire, Ramesh, Susana Ferreira, and Jeffrey H. Dorfman, "Flood-Induced Displacement and Civil Conflict," *World Development*, Vol. 66, February 2015.

Gleditsch, Nils Petter, "This Time Is Different! Or Is It? NeoMalthusians and Environmental Optimists in the Age of Climate Change," *Journal of Peace Research*, Vol. 58, No. 1, January 2021.

Gleditsch, Nils Petter, and Ragnhild Nordås, "Conflicting Messages? The IPCC on Conflict and Human Security," *Political Geography*, Vol. 43, November 2014.

Global Administrative Areas, "GADM Maps and Data," database, undated. As of May 22, 2023: https://gadm.org

Harari, Mariaflavia, and Eliana La Ferrara, "Conflict, Climate, and Cells: A Disaggregated Analysis," *Review of Economics and Statistics*, Vol. 100, No. 4, October 2018.

Hausfather, Zeke, and Glen P. Peters, "Emissions—The 'Business as Usual' Story Is Misleading," *Nature*, January 29, 2020.

Hegre, Håvard, and Nicholas Sambanis, "Sensitivity Analysis of Empirical Results on Civil War Onset," *Journal of Conflict Resolution*, Vol. 50, No. 4, August 2006.

Hegre, Håvard, Halvard Buhaug, Katherine V. Calvin, Jonas Nordkvelle, Stephanie T. Waldhoff, and Elisabeth Gilmore, "Forecasting Civil Conflict Along the Shared Socioeconomic Pathways," *Environmental Research Letters*, Vol. 11, No. 5, April 2016.

Helman, David, and Benjamin F. Zaitchik, "Temperature Anomalies Affect Violent Conflicts in African and Middle Eastern Warm Regions," *Global Environmental Change*, Vol. 63, July 2020.

Hendrix, Cullen S., and Stephan Haggard, "Global Food Prices, Regime Type, and Urban Unrest in the Developing World," *Journal of Peace Research*, Vol. 52, No. 2, March 2015.

Hoch, Jannis M., Sophie P. de Bruin, Halvard Buhaug, Nina Von Uexkull, Rens van Beek, and Niko Wanders, "Projecting Armed Conflict Risk in Africa Towards 2050 Along the SSP-RCP Scenarios: A Machine Learning Approach," *Environmental Research Letters*, Vol. 16, No. 12, December 2021.

Hoch, Jannis M., Sophie de Bruin, and Niko Wanders, "CoPro: A Data-Driven Modelling Framework for Conflict Risk Projections," *Journal of Open Source Software*, Vol. 6, No. 58, February 2021.

Högbladh, Stina, *UCDP Georeferenced Event Dataset Codebook Version 20.1*, Department of Peace and Conflict Research, Uppsala University, 2020.

Homer-Dixon, Thomas F., *Environment, Scarcity, and Violence*, Princeton University Press, 1999.

Ide, Tobias, "Rise or Recede? How Climate Disasters Affect Armed Conflict Intensity," *International Security*, Vol. 47, No. 4, Spring 2023.

Ide, Tobias, Anders Kristensen, and Henrikas Bartusevičius, "First Comes the River, Then Comes the Conflict? A Qualitative Comparative Analysis of Flood-Related Political Unrest," *Journal of Peace Research*, Vol. 58, No. 1, January 2021.

Inter-Sectoral Impact Model Intercomparison Project, "The Inter-Sectoral Impact Model Intercomparison Project," database, undated. As of August 8, 2023:
https://www.isimip.org/

ISIMIP—See Inter-Sectoral Impact Model Intercomparison Project.

Jones, Benjamin T., Eleonora Mattiacci, and Bear F. Braumoeller, "Food Scarcity and State Vulnerability: Unpacking the Link Between Climate Variability and Violent Unrest," *Journal of Peace Research*, Vol. 54, No. 3, May 2017.

Kahl, Colin H., *States, Scarcity, and Civil Strife in the Developing World*, Princeton University Press, 2006.

Koubi, Vally, "Climate Change and Conflict," *Annual Review of Political Science*, Vol. 22, May 2019.

Linke, Andrew M., Frank D. W. Witmer, and John O'Loughlin, "Weather Variability and Conflict Forecasts: Dynamic Human-Environment Interactions in Kenya," *Political Geography*, Vol. 92, January 2022.

Mack, Elizabeth A., Erin Bunting, James Herndon, Richard A. Marcantonio, Amanda Ross, and Andrew Zimmer, "Conflict and Its Relationship to Climate Variability in Sub-Saharan Africa," *Science of the Total Environment*, Vol. 775, June 25, 2021.

Martin-Shields, Charles P., and Wolfgang Stojetz, "Food Security and Conflict: Empirical Challenges and Future Opportunities for Research and Policy Making on Food Security and Conflict," *World Development*, Volume 119, July 2019.

Miro, Michelle E., Flannery Dolan, Karen M. Sudkamp, Jeffrey Martini, Karishma V. Patel, and Carlos Calvo Hernandez, *A Hotter and Drier Future Ahead: An Assessment of Climate Change in U.S. Central Command*, RAND Corporation, RR-A2338-1, 2023.

Muchlinski, David, David Siroky, Jingrui He, and Matthew Kocher, "Comparing Random Forest with Logistic Regression for Predicting Class-Imbalanced Civil War Onset Data," *Political Analysis*, Vol. 24, No. 1, Winter 2016.

Murakami, Daisuke, Takahiro Yoshida, and Yoshiki Yamagata, "Gridded GDP Projections Compatible with the Five SSPs (Shared Socioeconomic Pathways)," *Frontiers in Built Environment*, Vol. 7, October 2021.

National Intelligence Council, *National Intelligence Estimate: Climate Change and International Responses Increasing Challenges to US National Security Through 2040*, NIC-NIE-2021-10030-A, 2021.

Office of the Director of National Intelligence, *Annual Threat Assessment of the U.S. Intelligence Community*, February 6, 2023.

O'Neill, Brian C., Elmar Kriegler, Keywan Riahi, Kristie L. Ebi, Stephane Hallegatte, Timothy R. Carter, Ritu Mathur, and Detlef P. van Vuuren, "A New Scenario Framework for Climate Change Research: The Concept of Shared Socioeconomic Pathways," *Climatic Change*, Vol. 122, No. 3, February 2014.

Paul, Jomon Aliyas, and Aniruddha Bagchi, "Does Terrorism Increase After a Natural Disaster? An Analysis Based upon Property Damage," *Defence and Peace Economics*, Vol. 29, No. 4, July 2018.

Petrova, Kristina, Gudlaug Olafsdottir, Håvard Hegre, and Elisabeth A. Gilmore, "The 'Conflict Trap' Reduces Economic Growth in the Shared Socioeconomic Pathways," *Environmental Research Letters*, Vol. 18, No. 2, January 2023.

Pörtner, Hans-Otto, Debra C. Roberts, Melinda M. B. Tignor, Elvira Poloczanska, Katja Mintenbeck, Andrés Alegría, Marlies Craig, Stefanie Langsdorf, Sina Löschke, Vincent Möller, Andrew Okem, and Bardhyl Rama, eds., *Climate Change 2022: Impacts, Adaptation and Vulnerability, Contribution of Working Group II to the Sixth Assessment Report of the Intergovernmental Panel on Climate Change*, Intergovernmental Panel on Climate Change, Cambridge University Press, 2022.

Raddatz, Claudio, "The Wrath of God: Macroeconomic Costs of Natural Disasters," World Bank, Policy Research Working Paper 5039, September 2009.

Raleigh, Clionadh, Hyun Jin Choi, and Dominic Kniveton, "The Devil Is in the Details: An Investigation of the Relationships Between Conflict, Food Price and Climate Across Africa," *Global Environmental Change*, Vol. 32, May 2015.

Reuveny, Rafael, "Climate Change-Induced Migration and Violent Conflict," *Political Geography*, Vol. 26, No. 6, August 2007.

Riahi, Keywan, Detlef P. van Vuuren, Elmar Kriegler, Jae Edmonds, Brian C. O'Neill, Shinichiro Fujimori, Nico Bauer, Katherine Calvin, Rob Dellink, Oliver Fricko, Wolfgang Lutz, Alexander Popp, Jesus Crespo Cuaresma, Samir K. C., Marian Leimbach, Leiwen Jiang, Tom Kram, Shilpa Rao, Johannes Emmerling, Kristie Ebi, Tomoko Hasegawa, Petr Havlik, Florian Humpenöder, Lara Aleluia Da Silva, Steve Smith, Elke Stehfest, Valentina Bosetti, Jiyong Eom, David Gernaat, Toshihiko Masui, Joeri Rogelj, Jessica Strefler, Laurent Drouet, Volker Krey, Gunnar Luderer, Mathijs Harmsen, Kiyoshi Takahashi, Lavinia Baumstark, Jonathan C. Doelman, Mikiko Kainuma, Zbigniew Klimont, Giacomo Marangoni, Hermann Lotze-Campen, Michael Obersteiner, Andrzej Tabeau, and Massimo Tavoni, "The Shared Socioeconomic Pathways and Their Energy, Land Use, and Greenhouse Gas Emissions Implications: An Overview," *Global Environmental Change*, Vol. 42, January 2017.

Riahi, Keywan, Detlef P. van Vuuren, Elmar Kriegler, and Brian O'Neill, "The Shared Socio-Economic Pathways (SSPs): An Overview," poster, International Committee on New Integrated Climate Change Assessment Scenarios, undated.

Ru, Yating, Brian Blankespoor, Ulrike Wood-Sichra, Timothy S. Thomas, Liangzhi You, and Erwin Kalvelagen, "Estimating Local Agricultural Gross Domestic Product (AgGDP) Across the World," *Earth System Science Data*, Vol. 15, No. 3, March 2023.

Salami, Habibollah, Naser Shahnooshi, and Kenneth J. Thomson, "The Economic Impacts of Drought on the Economy of Iran: An Integration of Linear Programming and Macroeconometric Modelling Approaches," *Ecological Economics*, Vol. 68, No. 4, February 2009.

Schleussner, Carl-Friedrich, Jonathan F. Donges, Reik V. Donner, and Hans Joachim Schellnhuber, "Armed-Conflict Risks Enhanced by Climate-Related Disasters in Ethnically Fractionalized Countries," *Proceedings of the National Academy of Sciences*, Vol. 113, No. 33, July 2016.

Serneels, Pieter, and Marijke Verpoorten, "The Impact of Armed Conflict on Economic Performance: Evidence from Rwanda," *Journal of Conflict Resolution*, Vol. 59, No. 4, June 2015.

Sudkamp, Karen M., Elisa Yoshiara, Jeffrey Martini, Mohammad Ahmadi, Matthew Kubasak, Alexander Noyes, Alexandra Stark, Zohan Hasan Tariq, Ryan Haberman, and Erik E. Mueller, *Defense Planning Implications of Climate Change for U.S. Central Command*, RAND Corporation, RR-A2338-5, 2023.

Sundberg, Ralph, and Erik Melander, "Introducing the UCDP Georeferenced Event Dataset," *Journal of Peace Research*, Vol. 50, No. 4, July 2013.

Toukan, Mark, "International Politics by Other Means: External Sources of Civil War," *Journal of Peace Research*, Vol. 56, No. 6, November 2019.

van Vuuren, Detlef P., Jae Edmonds, Mikiko Kainuma, Keywan Riahi, Allison Thomson, Kathy Hibbard, George C. Hurtt, Tom Kram, Volker Krey, Jean-Francois Lamarque, Toshihiko Masui, Malte Meinshausen, Nebojsa Nakicenovic, Steven J. Smith, and Steven K. Rose, "The Representative Concentration Pathways: An Overview," *Climatic Change*, Vol. 109, August 2011.

van Vuuren, Detlef P., Elmar Kriegler, Brian C. O'Neill, Kristie L. Ebi, Keywan Riahi, Timothy R. Carter, Jae Edmonds, Stephane Hallegatte, Tom Kram, Ritu Mathur, and Harald Winkler, "A New Scenario Framework for Climate Change Research: Scenario Matrix Architecture," *Climatic Change*, Vol. 122, February 2014.

von Uexkull, Nina, Marco d'Errico, and Julius Jackson, "Drought, Resilience, and Support for Violence: Household Survey Evidence from DR Congo," *Journal of Conflict Resolution*, Vol. 64, No. 10, November 2020.

Witmer, Frank D. W., Andrew M. Linke, John O'Loughlin, Andrew Gettelman, and Arlene Laing, "Subnational Violent Conflict Forecasts for Sub-Saharan Africa, 2015–65, Using Climate-Sensitive Models," *Journal of Peace Research*, Vol. 54, No. 2, March 2017.

PHOTO CREDITS: Cover: A Yemeni soldier fights in the ranks of the legitimate army against the Houthi militia in Taiz City, Yemen, akram.alrasny/Adobe Stock; page viii: Rebuilding_Plan_Bee, Hashoo Foundation USA - Houston, TX (CC BY-SA 2.0, via Wikimedia Commons).